# THE DEATH OF THE SUN

*Also by John Gribbin*
*from Delacorte Press*

WHITE HOLES:
Cosmic Gushers in the Universe

TIMEWARPS

# THE DEATH OF THE SUN

## John Gribbin

Illustrated by Neil Hyslop

DELACORTE PRESS/ELEANOR FRIEDE

Published by
Delacorte Press/Eleanor Friede
1 Dag Hammarskjold Plaza
New York, N.Y. 10017

Manufactured in the United States of America

First printing

*Designed by Rhea Braunstein*

*Illustrated by Neil Hyslop*

Library of Congress Cataloging in Publication Data

Gribbin, John R
The death of the sun.

Bibliography: p. 183
Includes index.
1.  Sun—Popular works.  I.  Title.
QB521.4.G74        523.7          79–22191
ISBN 0–440–01924–9

For Douglas Orgill,
who *really* knows how to write

We have always wanted the Sun to be better than the other stars and better than it really is. We wanted it to be perfect, and when the telescope came along and showed us that it wasn't, we said, "At least it's constant." When we found it inconstant, we said, "At least it's regular." It now seems to be none of these; why we thought it should be when other stars are not says more about us than it does about the Sun.

DR. JOHN A. EDDY

Being stimulating can be more important than being right.

PROFESSOR MARTIN REES, FRS

# Contents

# THE DEATH OF THE SUN

# Introduction

THE DEATH OF THE SUN MIGHT SEEM A REMOTE PROBLEM OF little immediate human concern. After all, the best astrophysical evidence, described in this book, suggests that the Sun will continue pretty much as we see it today for another 4,000 or 5,000 million years. The theme is certainly a dramatic and important one—but why should we care about it here and now? And, indeed, why choose such a title for a book that is as much about the life of the Sun as its death? The reason is simple. Just because the Sun is, by human standards, so long lived, it has become a symbol of constancy for mankind, so that we speak, for example, of something being "as certain as that the Sun will rise tomorrow," while less than 400 years ago one of the great thinkers of the Renaissance, Galileo Galilei, suffered persecution at the hands of the religious authorities for daring to suggest, among other things, that the Sun might be imperfect. Against this background of the comforting familiarity of the Sun from day to day and year to year, and with still a deep-rooted feeling that this symbol of life on Earth is perfect and eternal, it is necessary to have a shock to the system before we can appreciate just how variable and imperfect the Sun actually

1

is. Although the day of its demise is a long way off, the fact is that the Sun *will* die; and with the realization of this reality we can also appreciate that the Sun must have been born and must change and evolve throughout its life, just as a baby changes and "evolves" on the way from the cradle to the grave.

So the death of the Sun is a key concept in appreciating that change is an essential feature of both the Sun and other stars. Once we accept that the Sun must die, it becomes easier to appreciate how smaller changes, on timescales important to life on Earth, happen throughout the Sun's life. Since Galileo's time, the struggle to understand such changes has continued, and generations of astronomers have argued, sometimes bitterly, over the evidence of solar variability. It is only in the 1970s that clear evidence has been gathered to convince all but the most doubtful of the doubting Thomases that solar variability not only happens, but is one of the key factors in molding conditions here on Earth. Astronomer Jack Eddy, one of the key figures in overturning the old view of solar constancy and ushering in the new era in which the Sun is seen as a rather seedy and variable star, has summed up the revolutionary nature of this change as follows:

There is good evidence that within the last millennium the Sun has been both considerably less active and considerably more active than we have seen it in the last 250 years. . . . The . . . implications of basic solar change may be but one more defeat in our long and losing battle to keep the Sun perfect, or, if not perfect, constant, and if inconstant, regular. Why we think the Sun should be any of these things when other stars are not is more a question for social than for physical science.

So, in a sense, something *has* died. The image of solar constancy, regularity, and perfection is dead—the "Sun" of popular mythology has died and been buried by modern science as surely as scientific studies dug the grave of the flat Earth theory many centuries ago. With this realization, we can open our minds to the new understanding of the Sun, and appreciate how the nature of the Sun and its changing relationship with planet Earth created the conditions under which life evolved, with solar changes often forcing changes in the environment of life on Earth. We can see how one day life on Earth will end as the Sun itself reaches the end of its life. And we can see how from decade to decade and century to century the Sun is even now tipping the balance of natural forces for or against mankind's continued role of top dog on Earth—while, just occasionally, even single dramatic outbursts of solar activity can produce immediate repercussions across space and shake us out of our complacent certainty that the Sun will be the same tomorrow as it is today.

JOHN GRIBBIN
April 1979

# CHAPTER 1

# Sundoom

THE TIME IS THE NEAR FUTURE. THE PLACE IS THE SHALLOW interior of our Sun, the surface skin a mere few thousand kilometers deep on a ball of hot gas 700,000 kilometers across, bigger than a million planet Earths. In these shallow outer regions energy is constantly transported outward from the interior, where temperatures reach 15 million degrees; the transfer process is convection, the constant stirring of material obeying the old dictum "hot fluid rises" that applies equally to a pan of boiling water or a hot air balloon.

But this stirring is not, in the case of our Sun, perfectly uniform. A variety of cosmic forces tug constantly at the Sun, pulling it first this way and then that. Magnetic fields coil like india rubber ropes around the charged particles of the solar gas, shifting them first one way and then the other. At times, when all the factors add up in the right way, the effects burst through to the surface and beyond, with spectacular results for mankind.

This is one of those times. For more than half a decade, magnetism has been building up as the magnetic ropes have been twisted in the solar interior, building toward the new peak of solar activity. With the whole convective zone trembling on the brink, another factor has lately been pushing in the same direction—the insistent tug of the planets in our Solar System, dominated by Jupiter and Saturn, swinging the whole Sun around through their gravitational pull and lining up to produce a jostle that comes only once or twice in a lifetime. Now the effects of this buildup of forces are about to be felt.

Deep in the convective zone a magnetic rope becomes twisted beyond its capacity, buckles and then breaks. Convection briefly runs riot as the magnetic forces blend into a new pattern, while a blast of superhot material shoots to the surface of the Sun, rising above in arching prominences, reaching half a million miles above the solar surface, and stretching at least as far in width, before being disrupted and hurtling into space. At the same time, lesser loops and fountains spatter the region of activity like a boiling mud swamp, while dark spots, marking the sites where magnetic ropes break through the solar surface and distort the normal convection pattern, blotch the solar disk.

At the heart of the sunspot group, a hot, explosive solar flare develops and blasts outward, producing a short-lived but powerful generation of radio waves, X rays, ultraviolet radiation, and bright visible light. Gases and shock waves pour out of the disturbed region, the most dynamic type of activity associated with the surface of the Sun. And this is the big one—the most powerful solar flare ever recorded, beating the spectaculars of 1959 and 1972 by a clear head.

By now the pattern of events is beginning to cause concern on Earth. A joint Nasa/European Space Agency shuttle mission has recently been launched, and is deploying space experiments; the flare alone is producing more X-radiation than the entire rest of the Sun, and particle radiation will follow not far behind. In space, unshielded by the atmosphere of the Earth, the dose of radiation may well prove lethal—no one has yet stayed in orbit through a major solar flare-up, and no one wants to be the first to find out whether the radiation is lethal or not. Energetic particles take only minutes to reach the Earth, traveling close to the speed of light—by the time we can see the flare, the radiation is already halfway to us. Fortunately, the earlier stirrings have not gone unheeded. Even before the flare itself burst into life the astronauts were shutting up shop, abandoning their mission, and attempting to return to Earth, and safety, in time. They may succeed—but even with the space shuttle it is touch and go whether reentry can be completed in the few minutes now left to them.

Halfway to the Sun, the impact of the radiation from the flare has already proved enough to knock out many of the sensors on a solar probe space vehicle, an unmanned instrumented spaceship designed to monitor any "normal" solar event and survive. But what is "normal"? The Sun has been around for 4,500 million years; we have been watching it with the aid of instruments for less than 400 years. What we think of as normal is turning out to be a rather narrow view. The concern among space scientists at the loss of their probe is almost balanced by their excitement in finding out that the Sun can produce something so spectacular and unexpected.

But not, perhaps, entirely unexpected. At least one group of physicists predicted more than three weeks in advance that something was about to happen, although even they missed the full significance of the coming flare. The Sun rotates roughly every twenty-eight days, and during the 1970s many pieces of evidence had been jigsawed together to link changes in the Earth's weather with solar activity four days earlier. The UK Appleton Laboratory had pioneered this study, and found that on some occasions distinctive "solar" influences on weather and geomagnetic patterns are found twenty-four days *before* a solar flare. In other words, the pattern occurs four days after the region of the Sun that was about to flare up last passed across the visible face of the Sun.

7

Just this evidence was used, three weeks before the present superflare, to forecast its occurrence—a warning was even sent to the European and American space agencies in an attempt to get the manned spaceflight postponed, but this was countered by the argument that the science of solar flare prediction was too vague to take seriously. As the shuttle returns now with millions of dollars worth of experiments ruined and the two astronauts who had been working outside the main space laboratory showing signs of radiation sickness, the egg on the faces of those agencies ensures that never again will such predictions be dismissed as "too vague."

For four days now the team that made the forecast has awaited the arrival of the slower moving particles from the gas hurtled outward in the flare-up, and the expected influence on low pressure weather systems developing at high latitudes. Meanwhile, the spectacular developments from the flare begin to provide entertainment for many and confusion for many more. Showers of charged particles plunge into the Earth's atmosphere, where they are funneled by the magnetic field toward the poles, cascading downward and interacting with other particles and the magnetic fields to produce bright aurorae, the northern and southern lights, on a scale never before witnessed in recorded history.

The Earth has its own layers of charged particles, high in the atmosphere, where the ionosphere is an invaluable aid to long-range communications, bouncing radio signals around the world. But not now. Disrupted by all this activity passing around and through it, the ionosphere is on the blink. A field day for some amateur radio enthusiasts, picking up signals they never hear under normal conditions; chaos for long-range broadcasters such as the World Service of the BBC, whose signals are either lost into space or bounced back to Earth in the wrong place. And overtime for the staff maintaining the hot lines between the superpowers, where heads of state maintain urgent contact, reassuring one another that no one is planning to start World War III by shooting off a few missiles while radar and communications are telling misleading tales.

But now the waiting is over for the solar-weather men. Out over Alaska, storms building up and coming in off the West Coast are beginning to deepen and swing east; in the Atlantic, south of Greenland, where the proximity to the magnetic pole makes the atmosphere particularly sensitive, the eastward moving depressions are boosted and kicked on their way toward England; at these latitudes, the spectacle of the bright northern lights is about to disappear behind a curtain of cloud, snow, and rain. The flare forecasters are euphoric—at long last, incontrovertible proof that the Sun's changing activity does directly affect the weather here on Earth. But still, this isn't the end of the story—far more than just the weather of planet Earth has been disrupted by this outburst from the powerhouse Sun.

A day after the intensive high-latitude storm activity that brought such joy to scientists at the Appleton Laboratory in England, routine monitoring of the spin rate of the Earth at the US Naval Observatory in Washington throws up another influence of the Sun on the Earth. As measured by superaccurate atomic clocks, compared with the passage of selected bright stars across the night sky, the length of the Earth's day jumps by a full tenth of a second between one night and the next. This is unprecedented, although rather smaller effects of the same kind were claimed by some scientists after the great flares of 1959 and 1972. What has happened is that the wind of particles from the Sun, blasting at the atmosphere of the Earth, has in effect made the planet "stagger" in its orbit around the Sun. Equilibrium will be restored—the "extra" tenth of a second will be "unwound" over the next week or so as the whole Earth, from the surface down to the molten magma of the interior, adjusts to this blow from outside. But this is not an easy process for anyone living on the surface of the planet—especially in regions prone to seismic activity.

Now, the problems of two sick astronauts and the loss of a few million dollars' worth of scientific equipment pale into insignificance. As the spinning Earth adjusts to the hammer blow from space, strain that has built up in the rocks for decades—in some cases centuries—is released thanks to this "last straw" effect. Volcanoes from Iceland to Antarctica and from Hawaii to Africa are stirred into rumbling life—not all of them, but far more than usual. Earthquake zones receive what seems in many cases to be an accustomed shaking; a ripple in the Middle East, spreading its measurable influence along the Mediterranean region; shakes in Japan and New Zealand; nothing out of the ordinary except that "normally" all three regions don't suffer the shakes simultaneously. Where major earthquakes have struck recently the effects are small, since little accumulated strain is waiting to be triggered. China, devastated in 1976, suffers hardly at all; but then the big one happens.

In California, earthquake attention has focused on San Francisco, in the north, where the last really big quake hit in 1906. But now it is the turn of Los Angeles, where the southern part of the state is about to be devastated by a disaster even bigger than the magnitude 8.3 earthquake that struck San Francisco then. Just north of Los Angeles, the ground is ruptured along a 300-mile zone, with a shift of more than seven meters horizontally from one side of the rupture to the other, and up to two meters in vertical displacement at various sites. The famous freeway system is in ruins; gas pipes and electrical cables rupture; and high in the mountains above the city dams built across the fault line crack, slide, and begin to crumble. First fire, fed by those ruptured gas mains, and then flood sweep across the devastated region, with evacuation hampered by the ruined road systems, which also prevent many emergency services reaching the stricken region. The death toll rises into five figures; the homeless into six; and the cost into tens of billions of dollars.

At last equilibrium is more or less restored. The Earth has shuddered under the solar hammer blow, shaken off the effects, and now proceeds in its orbit largely undisturbed, just as it has shrugged off the effect of similar blows many times in the past 4,500 million years. This minor incident in the history of the Solar System has, however, brought an almost crippling drain on the resources of the most powerful nation on Earth. And as recovery efforts get underway more grim news is at hand.

The coughing volcanoes, triggered like the great San Andreas earthquake by the direct results of one giant solar flare, have spread their burden of dust high into the stratosphere where it is being swept by strong winds around the world. The immediate effect is another spectacular display of color, this time in the form of brilliant sunsets that will last a decade or more. As it takes decades rather than months or years for the dust to settle, however, a more important consequence will make itself felt. High in the stratosphere, this volcanic dust blocks out some of the Sun's heat, preventing it from reaching the ground. The result is a slight global cooling—again, nothing spectacular in the history of the Earth, scarcely a degree Celsius overall. But it is enough to disrupt farming in Canada, North America, Europe, and Russia, bringing ten or more years of reduced yields at a time when the world's population is still increasing. Can this extra burden be accommodated, or will the world food system fall apart under the strain? What repercussions will there be in the political arena? As the world licks its wounds after a brief encounter with a solar flare, these are the problems for the more distant future.

This scenario is, of course, fictitious—in the sense that it hasn't happened yet. But it *could* happen, as soon as the early 1980s (next year, or the year after), when a rare alignment of the planets of the Solar System combines with a peak of sunspot cycle activity. Certainly such events have happened in the past, before mankind had a delicately balanced global civilization to worry about, and certainly they will happen again. The question is when—and how well can we cope with such kicks from the Sun? For the most dramatic discovery of modern astronomy—far more important to mankind even than the excitement of black holes—is the realization that our Sun may not be a "normal" star, at least at present. The evidence suggests that at the very least the Sun on which we depend is going through a period of instability, short-lived by the standards of stellar lifetimes (only a few tens of millions of years) and small on the stellar scale (only enough to change the temperature of the Earth by a few degrees either way). But that is more than enough for humankind, a race that has grown to world dominance *precisely* during the time the Sun has been "out of sorts." We are adapted to the Sun's being in this condition. What we are not fitted for is a return to its long-term "normal" state.

How much is the Sun likely to vary in the immediate future, now that we know that our beautiful theories of a perfect, unvarying Sun are wrong? When will it return to what is *really* normal—and will that be a good or bad thing for humanity? In a nutshell, can civilization cope with the crisis of the inconstant Sun? That is what this book is all about, and nothing which follows is fictionalized in any sense. Armed with this material, you can paint your own scenarios—but none of them are likely to be any less grim than the one outlined above.

CHAPTER 2

# Sun and Man

THE SUN IS SO VITAL TO EVERYTHING THAT LIVES ON PLANET Earth that it is taken for granted by almost everybody. Apart from a small amount of heat that leaks out from the interior of the Earth, and the influence of the Moon on tides, all of the energy around us, in both living and nonliving things, comes from the Sun, however devious the route may be. Atomic energy is something of a special case—not energy from *our* Sun, but bottled-up energy from events that took place long ago inside *other* stars, and this will be dealt with in its proper place. If the Sun "switched off" (which is, fortunately, something that stars like the Sun simply do not do!), then the Earth would be cold and dead. Energy from the Sun is the key to life as we know it.

Plants have evolved to trap the energy arriving from the Sun and to use it as the motive power of life; animals have evolved to steal energy from plants (or other animals) by eating them, but this makes them no less dependent on the Sun ultimately. So the story of man's dependence on the Sun must begin with the way plants trap that solar energy and store it in a form we can use.

The first step is to trap the energy and hold it in some kind

13

of a store where it is ready for use. Plants, and some bacteria, do this through photosynthesis—using the solar energy to make glucose (a form of sugar) and oxygen out of carbon dioxide and water; this is a chemical reaction that only works if energy is put into the system, and the glucose produced is richer in energy than the carbon dioxide and water it is made from.

Photosynthesis depends on the way packets of light energy from the Sun, called photons, can bounce electrons which are part of the molecules of the compound chlorophyll into energetic, or "excited" states. Molecules are made up from atoms of different substances, and the atoms themselves are made up of compact nuclei (sets of protons and neutrons bound together) set in a diffuse cloud of electrons. When the electrons are excited, they do the equivalent of jumping up a flight of steps by a fixed amount—the incoming energy shifts them to a more energetic state, balanced against the electric fields of the molecule. The situation is rather like the circus tumbling act where one person stands on a seesaw and two other people jump on the opposite end. The first tumbler is shot up into the air— raised to a higher energy state—by a fixed amount depending on the force of the jumpers on the other end of the seesaw.

Like the airborne tumbler, though, the excited electron in a molecule (chlorophyll in the case of photosynthesis) soon falls back to where it came from, unless something intervenes to stop it. With the tumbler, it may be that the extra energy is "used" to raise him high enough to land on someone's shoulders. In that case, the energy gained is stored until he jumps off, when it might be used, perhaps, to bounce a partner off a second seesaw. In a conceptually similar way, the electron raised to a high energy state by an incoming photon in a chlorophyll molecule can be captured before it has a chance to "fall back," when it would immediately give up its extra energy. Instead, it is caught in the high energy state and passed along a series of molecules called, logically enough, electron carriers. In effect, the Sun's energy has been turned into a tiny electrical current, as many electrons are excited and passed along the chain as long as the Sun shines.

This electric current is used to do two jobs. One is to break up water molecules into their constituent hydrogen and oxygen atoms, the other is to turn molecules of adenosine diphosphate (ADP) into a slightly more complex, and more energy rich, compound, adenosine triphosphate (ATP). The oxygen is not needed for the next step of photosynthesis, and is "thrown away" by plants, into the atmosphere. Hydrogen, carbon dioxide, and ATP, however, are the inputs for a series of chemical reactions which lead to the production of glucose, the object of the whole exercise; once the ATP is made, the trickle of electricity is not needed for this last phase of photosynthesis, which can take place in the dark, drawing on reserves of the required chemicals built up during the day.

All of this provides a store of energy. Animals, including man, are unable to tap sunlight directly through photosynthesis, but both plants and animals use the same basic method of tapping the store of energy and using it to drive the essential processes of life. This step is called respiration.

Both photosynthesis and respiration take place inside living cells, which are the factories of living creatures. In respiration, glucose produced by photosynthesis is broken down and its energy released; although plants make their own glucose and animals get theirs by eating some other creature's store, the process is the same. The first step in this process breaks down glucose into ATP and pyruvic acid; this doesn't release a lot of energy, but it has the advantage of working without any extra inputs—in particular, it doesn't require oxygen. Some living organisms go no further, having adapted to make do with the energy they get without adding oxygen; the most obvious example is the way yeasts ferment, by exactly this method, getting rid of the glucose by turning it eventually into alcohol and other

15

compounds. Muscle cells in animals can also do the same sort of thing for short periods when they are involved in strenuous exercise and can't get enough oxygen quickly enough; you can't get drunk on exercise, but the waste products do clog up the system and make the muscles "tired" so that they ache.

However, most plants and animals, for most of the time, take the process further. They use oxygen from the air to break down the molecules still further, forming new combinations and releasing a lot more energy. The chemical end products are carbon dioxide, released back into the air, and hydrogen—which still has another role to play. Hydrogen atoms without their electrons (in other words, positively charged hydrogen nuclei, which are single protons) go along a carrier chain like the chain which carries electrons in photosynthesis, with the electric currents used to make valuable ATP and the hydrogen eventually combining with oxygen to make water. The effect is that the food that we eat and air that we breathe are converted into carbon dioxide, water, and all-important energy. The picture is complicated by the ability of the body to store up energetic food in the form of fat, the protein of muscle, and so on; but this stored-up energy always comes from food, either plants we have eaten (bread, potatoes, or whatever) or animals which have themselves built up their food reserves by eating plants, especially grass.

Many bacteria do not use glucose as their primary food store, but other compounds such as ammonia; the story is still the same, however, in that the ultimate source of energy is the Sun. Even a microorganism that lives entirely on the waste products of the human body is dependent on us, just as we depend on plants. All it does is add another link to the chain stretching from the Sun's energy input across all living things on Earth.

There is a neat parallel between one thread in the process of photosynthesis and respiration and one of the many proposed "answers" to the world's present energy problems. The idea, put forward especially strongly by groups who are concerned about the dirtiness of present fuels and the dangers of radiation from nuclear power plants, is that solar energy could be turned into electricity which in turn could be used to break up ordinary seawater into hydrogen and oxygen, releasing oxygen into the air. The hydrogen could then be transported to where energy was needed and burned, combining again with oxygen and producing water to run back into the sea. Water broken up by electricity to make hydrogen; hydrogen transported to where it is needed and combined with oxygen to release energy and to produce water. Neat, clean, and efficient—and nature thought of it first, as is so often the case with neat, clean and efficient ideas!

At present, though, the idea of hydrogen as a major fuel in the world economy is still a pipe dream. The world today still depends overwhelmingly on fossil fuels—coal, oil, and natural gas—to keep the wheels of industry turning, automobiles running, and aircraft in flight. Where did the energy stored up in fossil fuels come from? Photosynthesis, the same process that builds up energy reserves in plants today, but carried out over millions of years, during which time the remains of dead plants and animals have been converted by geological processes into deposits that we regard today as fuel rather than food. Either way, their importance is the same; they contain stored-up energy, effectively bottled sunlight, which can be liberated under the right conditions by reactions with oxygen.

Coal has formed entirely from trees and other plants that lived their lives in swampy freshwater regions. When the vegetation died in such regions, its remains fell onto the swampy ground, building up over many generations. Each layer was buried by the next and the whole mass of dead vegetation was overlaid by sediments of nonorganic material washed down into the low lying swamps from surrounding higher ground. Sealed off from the air, and the oxygen which could otherwise have broken down the rotting remains until no food or fuel was left, the layers of vegetable matter became squeezed and transformed into peat, a dark brown or black fibrous material which still contains plant fragments and a lot of water.

Peat itself is a valuable fuel in many parts of the world. Layers tens of meters thick are found in many places at temperate latitudes, such as Ireland, and the deposits can be cut into blocks which dry out to form a fuel containing about 60 percent carbon, which burns to give carbon dioxide. Peat deposits are, however, relatively new by geological standards, being laid down in the past few tens of thousands, rather than millions, of years.

The very old deposits have been transformed further, being buried deeper under layers of sediment that turn to rock, being heated by the warmth inside the Earth at greater depths, and being squeezed by the geological forces that mold the rocks into new patterns over very long timescales. This process produces a continuous transformation, first into a soft, brown coal called lignite, then into successively harder and darker coals ending up with anthracite, very black and shiny with a carbon content of more than 90 percent, other constituents having been squeezed out or transformed along the way.

This process continues today. In different parts of the world all of the intermediate steps between living swamp vegetation and anthracite can be found. But because it takes many millions of years for the whole process to take place, there is no point in waiting for a known deposit of peat to turn into coal, or for a useful deposit of lignite to turn into the more valuable anthracite. On a human timescale, for all practical purposes we are stuck with a strictly limited amount of coal, the amount present in deposits under the ground now. New deposits will form, but they will be of use only to creatures as remote from us in the future as we are from the dinosaurs in the past.

Even so, the fixed amount of coal we have around now is

impressively large. Less than 2 percent of the world's coal has been used so far, and at least 8 million metric tons remain to be exploited, with vastly more buried too deep to be tapped by present-day technology but perhaps accessible to our descendants. Coal should serve mankind for hundreds of years— ample time to develop alternative energy supplies, such as the hydrogen cycle.

But the same cannot be said of oil and natural gas, which are in much shorter supply but in much greater demand.

Oil and gas account for well over two-thirds of world energy production today, but supplies are beginning to run out. The demand is great because the fuels are so convenient—liquids and gases can be piped around while solid coal has to be carted in trucks and shoveled into the fire, whether the fire is in a steam engine or a power station boiler. And the supply is low for the same reason—liquids and gases can only be trapped underground in very special circumstances, since usually they slip through holes and cracks in the rock and escape.

Oil and natural gas both derive from living organisms, as does coal, but they include animal as well as vegetable remains. (This makes no difference to the ultimate source of the energy; the animals survived by eating plants, or other animals which in turn ate plants, and it all comes back to photosynthesis in the end.) Unlike coal, though, oil and gas deposits are laid down mainly in coastal regions, near or under salt water. What happens is that marine organisms die and fall to the ocean floor, and in coastal regions, especially great river deltas, these deposits are supplemented by organic material washed down from the land by the rivers. Once again, the remains are buried by other remains, and eventually sealed off under sediments that build up to become new layers of rock. Over millions of years, the remains undergo similar chemical changes to those which produce coal, becoming converted in their turn to oil and natural gas. But unlike coal, the oil and gas move about through the rock once they form, flowing upward through porous rock layers away from the heat below and only stopping if trapped by a layer of impermeable rock (rock without any holes in it). Because rock layers are bent and twisted by geological forces, the oil and gas will accumulate under a dome-shaped region of impermeable rock, with the gas on top of the oil, held in place until someone drills a hole through the rock layers and lets the fuel out.

But most of the oil and gas never got trapped in this way, and the deposits we find today are a small proportion of all the animal and vegetable material that has ever been turned into oil and gas in the rock layers beneath the ground.

It's not only the oil and gas that move; the rocks themselves, in addition to being squeezed and distorted, are physically moved around the globe through the mechanisms of continental drift,* so that oil and gas are not found exclusively in coastal regions today. Indeed, some oil is found in high mountain regions, such as the northern Andes of South America. But wherever oil and gas are found, there once was a layer of rock beneath the ocean. In the same way, coal is found today in regions far away from tropical swampland, such as the plains of Poland.

Various intermediate forms of organic remains exist in different parts of the world, where carbon-rich material may be locked up in finely grained rock as oil shale or in sandstone as tar sand. All of the energy in all of these reserves, though, came originally from the Sun. So there is a neat, if unhelpful, answer to anyone who urges us to use solar power—we already do! Even burning wood for fuel is using solar power, stored up by the tree which used photosynthesis to provide the energy to build up its carbon-rich bulk, stealing carbon from carbon dioxide in the air and releasing oxygen. Small wonder, then, that one

*See my book *Our Changing Planet.*

20

of the most attractive potential solutions to the rate at which the Earth's limited oil reserves are being depleted is to farm specially chosen crops, such as sugar, which can be easily converted to alcohol to be burned in place of gasoline. All we are doing then is cutting out the geological middleman, which may be efficient in energy use but which takes too many millions of years to do the job for any human's patience.

And if the wheels of our civilization are lubricated by oil, which itself is bottled sunlight, can any modern person feel that our ancestors (and our contemporaries in some allegedly backward regions of the globe) made any mistake in their priorities by worshiping the Sun as a god—not just "a god," but the most important god of all, to those cultures who believed in the existence of more than one.

Of course, the power of the Sun over man was not always acknowledged, even then. It is too disturbing to believe that we may be entirely at the mercy of a power beyond our control, so that much of Sun worship is concerned with a belief that by undertaking certain rituals man is *controlling* the Sun. This ranges from the tribes who shoot arrows skyward during an eclipse to frighten off the monster that is devouring the Sun— a method which always works!—to offerings made every morning to ensure that the Sun will rise and provide his life-giving light and warmth during the day.

Clearly the links between Sun and life were apparent to any developing human culture. But this led in many cases to strange rituals and taboos developing from fear that the Sun's life-giving power may be too powerful—too magic—on occasions. There are many regions of the world, for example, where cultures developed taboos limiting the extent to which girls at puberty were allowed to expose themselves to the Sun (or, indeed, limiting the extent to which they were allowed to touch the ground, home of that other power, the Earth God). Many examples are cited by J. G. Frazer in his classic study of magic

21

and religion;* one is of the Indians of California, who thought a girl at her first menstruation was "possessed of a particular degree of supernatural power . . . an attempt was made to seclude the world from her . . . she kept her head bowed and was forbidden to see the world and the sun"; another example cited is the Tsetsaut tribe of British Columbia, who believed that if a girl at puberty exposed her face to the Sun, then rain would fall. Given the climate of British Columbia, this seems almost as certain as the fact that the Sun will reappear after an eclipse if we shoot arrows at it!

Perhaps the most important influence of the Sun on religions, though, stems from its own rhythmic cycles. The Sun leaves us every night, only to return reborn the next day, and it rises high in the sky in summer, slips further away from us in winter, and is rejuvenated in the spring. Small wonder that concepts of death and resurrection, life after death, reincarnation, and the like permeate the religions of mankind. And as a reminder that we should not be too superior about how our modern scientific knowledge transcends these primitive beliefs, it is salutary to look at one of the most important dates celebrated as a holiday in the developed, Western world.

The date is December 25, Christmas Day. Most people appreciate that Jesus Christ was not born on December 25, and that this date has simply been chosen as a convenient one on which to celebrate his anniversary. But why was December 25 chosen in the first place by the leaders of the early Christian Church?

At the time Christianity became established, there were many competing religions in the then civilized (that is, the Roman) world. Among these, one of the most popular was the cult of Mithra, which has points of resemblance with Christian teaching, and was seen as a particularly dangerous opponent by

*J. G. Frazer, *The Golden Bough;* see bibliography.

22

early Christians—indeed, as the work of the Devil, a false religion enough like Christianity to trap the unwary and seduce them away from the true faith. Both religions combined solemn rituals with teachings of moral purity and a prospect of immortality,* and there was a long struggle between them for supremacy. Christianity won out in the end, but not without being changed in the process.

Mithra was identified by his worshipers with the Sun, and to Sun worshipers one of the most important dates in the calendar is the winter solstice, the shortest day when the Sun sinks lowest on the horizon. After the winter solstice, the Sun climbs back to full power and strength, and we can see the signs that spring is on the way by the gradual lengthening of the day and decline in the hours of darkness. Quite naturally, for Sun worshipers, the winter solstice can be identified with the day of birth, or rebirth, of the Sun. When is the winter solstice? On the modern calendar it falls on December 22. But according to the Julian calendar used by Rome in the centuries after Christ, and within the then-known accuracy of astronomical observations, the winter solstice was reckoned to be December 25.

According to Frazer, writing in *Adonis, Attis, Osiris,* after the Eastern Christian Church had got into the habit of commemorating the birth of Christ on January 6 it took until the late fourth century for the Western Church to mark the festival at all, and at about that time both branches of the Church fixed on a new date, December 25. Why did they change? Frazer quotes the writings of a Syrian Christian of the period:

The reason why the fathers transferred the celebration of the sixth of January to the twenty-fifth of December was this. It was a custom of the heathen to celebrate on the same twenty-fifth of December the birthday of the Sun, at

*See J. G. Frazer, *Adonis, Attis, Osiris.*

23

which they kindled lights in token of festivity. In these solemnities and festivities the Christians also took part. Accordingly, when the doctors of the Church perceived that the Christians had a leaning to this festival, they took counsel and resolved that the true Nativity should be solemnized on that day.

Our modern Christmas festival, lights and all, is a direct descendant of the pagan festivals to commemorate the winter solstice and mark the rebirth of the Sun—the lights themselves are sympathetic magic, supposed to help the Sun regain his strength and brightness. Down the centuries the echoes of this have disturbed various religious authorities, including Leo the Great, who commented on the belief that Christmas marked the birth of the new Sun, not the birth of Christ. Today, the issue seldom raises passions in any breast, and the festival of Christmas is celebrated by many people who are not Christians. The wheel has come almost full circle, and we are back to a midwinter festival celebrated chiefly to cast off the gloom of the long winter nights.

People certainly do feel an affinity for the Sun, marked by the annual rush to sunny climes for holidays and, on the blacker side, by the increased incidence of suicides during the bleak month of January (in the northern hemisphere). And we have seen how vital the Sun is to the continuation of life on our planet. But where did the material of our planet, and of living things, come from in the first place? Not from our Sun; as I have hinted already, everything on Earth, and the Earth itself, has been processed through the inside of at least one star before coming together to form the living and nonliving surroundings we see about us.

Stars glow because they are hot, and they are hot because of nuclear reactions going on inside them. In these reactions, simple atomic nuclei are fused together to form more complicated nuclei—lighter elements are fused to make heavier elements. In the process, a little mass is turned into energy, following Einstein's equation $E = mc^2$, and this provides—adding the energy from billions of fusion reactions—the heat which makes the star glow. We shall see later the importance of this for the life of a star like the Sun; here I want to stress that we are all, literally, stardust.

The process by which elements are built up in nuclear reactions is called nucleosynthesis. When the Universe itself was born in a great outburst some 20,000 million years ago (the "Big Bang"), the matter that was created was chiefly hydrogen, the simplest element of all, with a little helium, the next simplest element. Everything else has been made inside stars by nucleosynthesis. So the first stars that were made contained only hydrogen and helium; even today, hydrogen is by far the dominant material of the Universe, making up 92 percent of all matter. Helium makes up almost all of the other 8 percent; everything else is, on the scale of the Universe, a very minor trace of impurities. Living as we do on a rocky planet, with bodies made up of elements like carbon and breathing oxygen from the air, this is hard to appreciate. But even within our Solar System the Sun itself is so much bigger than all the planets put together, and so rich in hydrogen, that overall the heavier elements are very much in the minority. And there are great clouds of hydrogen gas between the stars, to be taken into account when working out the chemical element budget of the whole Galaxy of stars, the Milky Way, in which the Solar System is but a tiny part.

Inside a star, hydrogen is converted into helium in the first stage of stellar nucleosynthesis. This takes place at a temperature equivalent to 10 million degrees ($10^7$ K), until all the hydrogen in the core is used up. Then, the star contracts and warms up in the center to about 100 million degrees ($10^8$ K), when the helium nuclei fuse together to form the elements carbon and oxygen, releasing more energy in the process. Further steps in the chain see carbon and oxygen nuclei involved in fusion reactions building up heavier elements still, such as neon, sodium, magnesium, and silicon. All of these steps up the ladder involve elements whose nuclei have a mass which is a multiple of four times the mass of the elementary particle called the proton. This is because helium nuclei, each with a mass of four units, are the basic building blocks of the process. Eventually, the chain builds up to produce iron, which has a nuclear mass of 56 units, equivalent to 14 nuclei of helium-4.

Once the process reaches this peak, other elements begin to be produced, filling in the gaps, at temperatures corresponding to 4,000 million degrees ($4 \times 10^9$ K). The energies involved in such a hot stellar core with many particles colliding with one

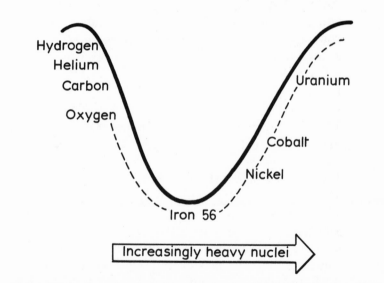

*Figure 1* The iron-56 valley. From either side of the valley, energy can be liberated by moving toward iron-56. Light elements fuse together, releasing energy, but heavy elements release energy by fission.

26

another cause individual protons and helium nuclei (also known as alpha particles) to be knocked out of the nuclei until a whole range of masses—corresponding to a whole range of chemical elements—is built up. But there the fusion process ends. Elements heavier than iron-56 cannot be made by fusion in the same way, because instead of liberating energy the addition of more nuclear particles to make heavier elements requires energy to be put in.

It's as if iron-56 lay at the bottom of a valley, between mountains representing the lighter elements on one side and the heavier elements on the other. From the light side, energy is released by "rolling down" toward iron-56, making heavier elements on the way. But from the heavy side, to release energy and get into the iron-56 valley, the heavy atomic nuclei must be broken up. This is called fission, by contrast to the fusion process. So where do the heavy elements (heavier than iron-56) come from?

The iron-56 nuclei can, in fact, pick up extra particles—neutrons, the neutral counterparts of the positively charged protons—from the sea of energetic particles inside a hot stellar core. This undoubtedly explains the formation of some of the heavier elements. Now the process is taking energy out of the core, effectively cooling the star down in the process. But a great deal of energy is needed to make measurable quantities of the heaviest elements, and it is not absolutely clear just where this comes from. Most probably, these elements are produced when stars explode as novae or supernovae; just possibly, some were produced, like helium, in the Big Bang. But ultimately the energy which has been used to drive up into the heavy element mountain out of the iron-56 valley is borrowed energy, from the fusion of lighter elements. It is as if you rode on a bicycle down from the light element mountain and across the valley, using the momentum to carry you high up into heavy element mountain. It is only because there is so much hydrogen in the Uni-

27

verse that stars can use this vast store of energy to glow brightly for thousands of millions of years and still have a bonus left over to make very heavy elements. And it is because of the energy needed to make them that elements heavier than iron are rare, with very heavy elements being very rare.

Among those very heavy elements, though, are those with radioactive nuclei such as uranium and plutonium. These elements can be encouraged to fission quite simply, giving up energy as they hurry back down toward the stable iron-56 valley. This fission process is the basis of all nuclear reactors yet built—so for atomic energy we are dependent, not on the Sun, but on other stars that built up heavy elements and scattered them into space to become the stuff of the Solar System in which we live.

In nuclear fusion, however, the so-called "hydrogen bomb" draws on no reserves of bottled stellar energy. It is literally a star in miniature—and when fusion power can be tamed inside a steadily running reactor, then and only then will mankind have access at last to the energy of the stars under his control. Meanwhile, we turn to the Sun for life itself, and remember that everything about us, including our own bodies, has been precooked inside other stars. The living inhabitants of planet Earth are children of the stars and ruled by the Sun; but what of the physical links between the Sun and the Earth itself?

CHAPTER 3

# Sun and Earth

LIFE DEPENDS DIRECTLY ON THE SUN, AND MAN DEPENDS ON stored solar energy to keep his technological civilization going. But the Sun's influence extends even further, molding our whole environment here on planet Earth. Going right back to basics, the Earth simply would not be the kind of planet it is if the Sun weren't the kind of star it is.

Our Solar System contains two kinds of planets and some bits and pieces of cosmic junk (I include Pluto, the outermost "planet," in the junk category, since it is almost certainly an escaped moon from one of the giant planets). Orbiting relatively close to the Sun there are four rocky planets, Mercury, Venus, Earth, and Mars; these are sometimes called the "terrestrial" planets, not because they are all *exactly* like the Earth but because, like the Earth, they are all solid planets with only thin layers of atmosphere (in the case of Mercury, closest to the Sun, scarcely any atmosphere at all). Further out from the Sun, past the main collection of junk in the asteroid belt, there are four giant planets, Jupiter, Saturn, Uranus, and Neptune. These are quite different from the terrestrial planets.

For a start, they are much bigger. Jupiter, the biggest, has

29

two and a half times as much mass as all the other planets in the Solar System put together, and a volume equal to 1,319 Earths; even Uranus and Neptune, the two smallest gas giants, are about 60–70 times the volume of our home planet. As well as sheer size, the giants are very different in composition. They are mainly gaseous, possibly with small rocky cores, and the gases of which they are composed are rich in hydrogen-compounds like methane and ammonia.

There is one underlying reason for both these differences between the two kinds of planet—which, it turns out, are really two facets of the same difference, the key to which is the heat of the Sun. When planets formed out of the dusty material of the solar nebula, the cloud of gas and dust from which the Sun itself condensed, they did so in the face of conflicting forces. At different distances from the Sun, where lumps or aggregations of matter began to get together, the gravity of each lump would tend to pull them into bigger congregations, slowly building up to form planets. But at the same time, the hot young star at the middle of the young Solar System was warming the material up, making the light elements blow away into space. The heat was greatest nearest the Sun, of course, so very light gases stood almost literally a snowball's chance in hell of sticking to the planets forming closest to the Sun.

Further out, where things were, and are, cooler, the Sun's heating effect could be overwhelmed by the small but insistent tug of gravity, holding even atoms as light as those of hydrogen in place around the new planets.

Hydrogen, as we have seen, is the lightest element, as well as being the most common in the Universe, the basic building block from which other elements are made by fusion. So almost all the hydrogen blew away from the inner Solar System as the planets formed. The terrestrial planets are made up of leftovers, after almost all of the most abundant elements, hydrogen and helium, were blown away in the hot solar wind; the Earth is made up of the less abundant elements (which constitute less than 1 percent of the matter of the Solar System), with just a little hydrogen that was trapped in chemical compounds of sufficiently heavy molecules. This little bit of hydrogen is, however, very important to us—it is locked up chiefly in water molecules today, and we have already seen what an important part hydrogen plays in the chemistry of life.

The outer, giant planets are much more typical of the composition of the whole Universe, or even of the Sun, as they contain a great deal of hydrogen. But even they have lost a lot of hydrogen, since the gas is light enough to escape from the atmosphere of even a planet as big as Jupiter. A lot of hydrogen remains in those heavy molecules such as methane and ammonia—but after all, in the Sun, or in those great clouds of cool gas between the stars, there is mainly pure hydrogen, without the less common elements such as carbon and nitrogen (the central atoms of methane and ammonia), except as a trace.

Getting back to the inner Solar System, even the four terrestrial planets show differences caused by the Sun's heat, an influence which has persisted from their birth to the present day. Mercury, the innermost planet, which formed under the hottest conditions, is very dense and has a relatively large core probably composed of metals. Just as the inner planets are in a sense like the cores of giant planets stripped of their hydrogen-rich atmospheres by solar heat, so Mercury is in some ways like the core of a genuinely terrestrial planet, stripped even of the elements that might have made a rocky crust. Venus and Earth, the next two planets out from the Sun, are almost the same size as each other and have similar compositions, while Mars, further out still, is an even lighter planet, deficient in metals by the standards of planet Earth.

So the nature of the rocks beneath our feet was determined critically by the heat of the Sun when the Earth was forming, and by the distance of that forming planet from the Sun, the

31

radius of the Earth's orbit. A little closer and there might not have been much in the way of rocky crust at all; a little further out, and the rocky Earth might have been merely a buried core deep below the thick atmosphere of a gas giant. And just like the rocks beneath our feet, the nature of the air that we breathe depends crucially on the heat of the Sun and our distance from it.

Mercury has no atmosphere to speak of; Mars has a thin atmosphere, but is too cold for water to exist as a liquid or for rain to fall. Venus, so nearly Earth's twin in many ways, has an atmosphere that is *too* thick, a blanket of carbon dioxide which traps solar heat through the "greenhouse effect," raising the temperature at the surface of the planet to nearly 500°C. The Earth alone has an atmosphere sufficient to keep temperatures at the surface between the boiling point and the freezing point of water in most places for most of the time. The result is a planet rich in water, where water vapor evaporated from the oceans is returned to the surface as rain, and this is ideal for life as we know it.

Just possibly, there may be life forms on Mars or Venus, but only on Earth has life got a firm grip on the surface of a planet in our Solar System, spreading into every possible ecological niche—and that profusion of life is intimately connected with the profusion of water. Our very word for a region devoid of life—desert—is synonymous with a region deficient in liquid water. So how did the Earth get to be such a green and pleasant place—and, in particular, what part did the warmth of the Sun play in the evolution of the Earth's atmosphere?

The only handle which atmospheric scientists have on the problem is the average surface temperature of a rocky body at the distance of the Earth (or Mars, or Venus) from the Sun. Fortunately, though, this turns out to be just the handle they need! The experts generally agree that all three planets lost any original, or "primordial," atmosphere very early in the history of the Solar System, when the young Sun reached a peak heat before settling down into steady nuclear burning. Their present atmospheres are the result of gases escaping from the interior, including both steady and explosive "outgassing" from volcanic activity, and the sudden vaporizations produced by the impacts of large meteorites. Today, such outgassing produces mainly water and carbon dioxide.

It used to be thought—and is still widely taught—that the first stage of outgassing produced a mixture of methane and ammonia, gases like those found in the atmospheres of the giant planets. But there is no sound physical reason for this belief; a rocky planet that produces water and carbon dioxide by outgassing today is much more likely to have produced water and carbon dioxide by outgassing when it was young. The reason why ammonia/methane atmospheres were fashionable, though, is easy to find. Quite simply, the atmospheric scientists were trying to help their biological colleagues by providing an atmosphere suitable for the origin of life.

The origin of life on Earth is one of the key puzzles of science, and several decades ago scientists found that by mixing gases like methane and ammonia with water in sealed tubes, and passing electric sparks or ultraviolet light through the mixture, they could make quite complex organic molecules, the precursors of life. The early Earth certainly had lots of water, plus electric sparks (lightning) and ultraviolet radiation from the Sun, so the methane/ammonia early atmosphere was added by the theorists, essentially through wishful thinking. More recently, though, other experiments have shown that the precursors of life can build up through chemical reactions even in atmospheres which are chiefly carbon dioxide, and there is a dramatic new theory—discussed in more detail in Chapter 4—which suggests that these complex molecules can even form in clouds of interstellar material, ready to "seed" young planets at a later date. Whichever idea is correct, there is no longer

33

any need, in terms of the origin of life, to postulate a methane/ammonia atmosphere at any time in the evolution of the Earth. We are free to make the simplest and most logical assumption, that volcanic outgassing has always produced the same types of gas, and that from the very beginning, some 4,500 million years ago, the Earth's atmosphere—and those of Mars and Venus—were mainly made of water vapor and carbon dioxide. How, then, did the three planets get such different atmospheres today?

Venus and Mars, in fact, still *have* carbon dioxide atmospheres, although one is thick and hot and the other is thin and cold. Nothing much to explain there, and everything that does need explaining fits in well with the positions of the orbits of those planets. For a rocky planet at a known distance from the Sun, there is one stable temperature which balances the heat coming in from the Sun and the heat being radiated away into space by the hot rocks. For Venus, this temperature is about 87° C, and that must have been the surface temperature when the first atmosphere began to form as gas escaped from the rocks. That temperature was high enough so that right from the start all the water released stayed as vapor, acting with carbon dioxide to trap heat near the surface.

Both carbon dioxide and water vapor trap infrared heat, which is the kind radiated by hot rocks (or by the warm glow of an electric fire). But they are both transparent to the kind of heat coming in from the Sun, which contains very little infrared radiation. So they act as a blanket around a planet, tipping the balance between incoming radiation and outgoing radiation so that the surface heats up. This is the "greenhouse effect." So, for Venus, the initial temperature of 87°C was just the start—as more and more gas was released, the surface got hotter and hotter, soon rising above even the *boiling* point of water and producing the hothouse we see today.

On Mars, things were very different, with a stable surface

temperature of almost 30 degrees *below* zero before outgassing got underway. Water could not even melt, let alone evaporate, and the result was a thin atmosphere with water frozen below the surface, which we still see today. It is just possible, incidentally, that Mars has experienced more exciting times. Parts of the surface show very clear signs of erosion by what might have been running water, with dried-up riverbeds, canyons, and so on. Calculations show that with only a slightly thicker atmosphere Mars would reach a warmer state, thanks to the greenhouse effect, which could keep water liquid, even though conditions would still not be right for rainfall, lakes, and oceans like those on Earth. A slight change in the tilt of the planet toward the Sun could have evaporated the polar caps, perhaps giving enough new atmosphere to do the job, or perhaps the impact of a giant meteorite could have melted and evaporated frozen water near the surface, doing the same job. Either way, the situation certainly was not permanent, and Mars cooled off as the atmosphere itself slowly froze and thinned out—as far as we can tell, judging from the number of meteorite craters that scar the "rivers" of Mars, it is at least 500 million years since water, or whatever liquid it was, flowed on our neighboring planet.

Earth remains the odd planet out, between the extremes of heat and cold, providing a comfortable home for life. Here, the initial surface temperature was about 25°C, enough for liquid water to flow but not so high that enormous quantities of water vapor got into the atmosphere to produce a runaway greenhouse effect. Quite the reverse—the warm waters dissolved carbon dioxide out of the atmosphere, checking the greenhouse effect further and actually cooling the planet a little to an average of around 15°C. This is because, with an atmosphere, especially one containing clouds of water vapor, some of the incoming solar heat is reflected away before it ever gets to the surface of land or sea.

For a planet such as Earth, at our distance from the Sun,

35

there is even a kind of inbuilt thermostat to keep the temperature close to 15°C. Suppose the Sun warmed up a little, as it might during its lifetime. Then, instead of the Earth getting hotter and perhaps developing a runaway greenhouse effect, the slight increase in temperature might produce more evaporation from the oceans, with more clouds, which would reflect away solar heat and limit the rise in temperature. Or imagine a slight cooling; with less evaporation and fewer clouds, more solar heat could reach the ground, making the cooling less severe than it would otherwise have been.

These are just hypothetical examples. The important point is that the atmosphere and oceans of our planet have evolved in close harmony with the heat arriving from the Sun. With a slightly cooler or hotter central star from the outset, and the same distribution of planets, it could have been Venus or Mars that were ideal for life and Earth that was either baked or frozen.

Life itself now plays a part in the atmospheric balance, and there are two schools of thought about this. One is the "Gaia" hypothesis, put forward by British Professor Jim Lovelock, who sees life acting as a regulator on the atmosphere, producing just the right balance of heat-retaining gases to keep conditions ideal for the life forms that are making the gases. The other view is that mankind may upset the applecart, especially by producing so much carbon dioxide from burning fossil fuel that a runaway greenhouse becomes established. Remember, though, that all of the carbon in fossil fuel was once in the atmosphere as carbon dioxide, and didn't produce a runaway greenhouse effect then; and, lest we think man may now be rivaling the Sun as an influence on the Earth's atmosphere, never forget that man's energy, both food and fuel, is itself transformed solar energy. Anything *we* do to our planet is also a solar effect—and if it is photosynthesis that maintains the balance of gases so comfortably for life on Earth, that too is an indirect influence of solar energy, acting through the living processes of plants.

Obviously, if something went wrong with the Sun we would be in trouble in a big way. But even with a perfectly steady, well-behaved Sun changes occur on Earth because of solar heating effects. Remember that the present balance is struck between the Sun's heat and the Earth's position in space; even if the Sun's heat doesn't vary, the Earth's position in space certainly does, as it orbits around the Sun. And the orbit itself also changes over tens of thousands of years. These effects bring us the changing pattern of the seasons and, climatologists now believe, explain the detailed fluctuations of temperature in ice ages.

The differences between winter and summer are certainly

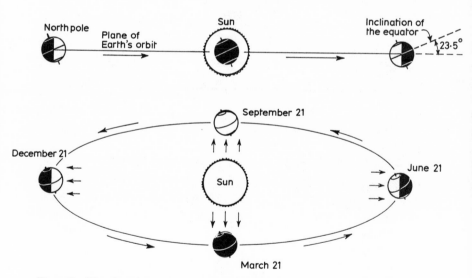

*Figure 2* The changing seasons result from the way the Earth's axis (the line joining the North and South poles, on which the Earth spins) is tilted so that the plane of the equator makes an angle of 23.5° with the plane of the ecliptic (the plane of the Earth's orbit around the Sun).

profound for anyone living at middle or high latitudes. One neat, and apt, description is that each year the high latitudes of the Earth experience the start of a new ice age, but fortunately each year this is soon replaced by spring and then summer. The reason is simple—and has nothing to do with any changes in the output of the Sun.

Because the Earth is tilted at an angle of about 23.5 degrees away from the perpendicular to a line joining the center of the Earth and the center of the Sun, days are not exactly divided into twelve hours light and twelve hours darkness, and the pattern of solar heating over the Earth's surface changes over the year. In December, the northern hemisphere is tilted away from the Sun, so that the nights are long. At the same time,

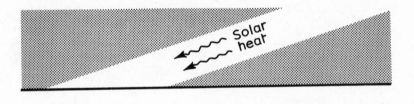

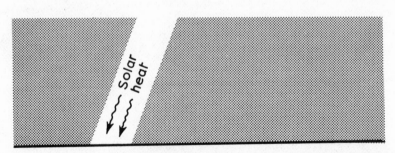

*Figure 3* *Top:* With the Sun low on the horizon (winter) heat spreads thinly over the ground. *Bottom:* With the Sun high in the sky (summer) the same amount of heat is concentrated in a smaller area on the ground.

38

because of the tilt, the radiation from the Sun is spread more thinly over the ground, from a Sun seen low on the horizon. Meanwhile, in the southern hemisphere the reverse is happening—short nights and long, warm days with the Sun high in the sky. Six months later, the positions have been reversed, because although the whole Earth moves around the Sun its direction of tilt stays more or less the same. This, in a nutshell, is the cycle of the seasons, with the intermediate spring and fall occurring when the Earth is "sideways on" to the Sun in its orbit.

But this cycle is not exactly fixed—the "more or less" constancy is just that, and the extent of the changes is sufficient to tip the whole Earth into or out of an ice age proper. First of all, the Earth's orbit is not a perfect circle but an ellipse, so that we are actually a little bit further away from the Sun during northern summers, at present, than during northern winters. The season of closest approach to the Sun (perihelion) varies slowly, but in a regular, predictable way, over tens of thousands of years. In addition, the amount of the tilt of the Earth away from the perpendicular varies, also in a regular way, and the direction in which the tilted Earth "points" follows its own cyclic variation. The net result of all these cyclic changes is that the season of closest approach to the Sun varies, the extent of the difference between the seasons varies, and the pattern of heat variations over the surface of the Earth changes. Without any change in the total amount of heat received in each year over the *whole* surface of the Earth, this alone is sufficient to take us in and out of ice ages.

In a nutshell, when we have cool summers in the northern hemisphere and very cold winters in the southern hemisphere (conditions which, of course, go hand in hand), some of the winter snows in the north fail to melt each year, building up into ice sheets, while each year more and more sea ice freezes in the south, so that glaciation develops at both extremes to-

39

gether.* Geological studies show beyond question that this subtle rhythm of orbital and tilt variations explains precisely the recent changes in snow and ice cover of the globe—but by "recent" we mean the past couple of hundred thousand years or so, and we mean broad changes in climate on the scale of ice ages, not subtleties like last winter's weather. These effects work too slowly, and too subtly, to have any influence in a human lifetime. Nevertheless, we do have a clear forecast of where the Earth is going, in terms of climate, from projecting these astronomical changes forward in time.

It seems that the relative warmth the Earth has enjoyed since the end of the most recent ice age, no more than fifteen thousand years ago, is just about over, and we are heading back into a situation of cooler northern summers and vicious southern winters. Although the effects are too small to be measured in a human lifetime, they will build up until, within five thousand years, and perhaps in only half that time or less, the Earth will be well on the way back into a full ice age, from which it is unlikely to emerge for a further hundred thousand years, other things being equal. This is typical of the pattern of recent times —relatively warm "interglacials" only a little over ten thousand years long, separating much longer ice ages of a hundred thousand years or so. All of human civilization has developed in one of these warm breathing spells, a minor interglacial ripple in a succession of ice ages. In terms of human civilization, rather than a single human lifetime, this is the sort of timescale we have to consider. Perhaps, given a thousand years or more, we may develop the technology to cope with a new ice age. But then again, although ice ages are "normal" in terms of the past few million years of the Earth's history, looking even further back we find much warmer conditions for many more millions of years—a long-term pattern in which several hundred million

*For a detailed discussion, see my book *What's Wrong With Our Weather.*

years of warmth is followed by about ten million years of cold (an ice epoch) and so on. Perhaps we should worry just as much about the prospect of a return from a dominant icy pattern into a very warm pattern, with melting ice caps and rising sea levels providing the long-term problems.

There are reasons to be found on Earth for part, at least, of these changes. The orbital effects account for a lot of the climatic variation between ice ages and interglacials; and geophysicists now tell us that because the continents move about the face of the Earth, on a timescale of hundreds of millions of years they will only occasionally arrange themselves into a pattern where ice ages are possible at all, with plenty of land near the poles on which snow can fall and build up into ice sheets.* This may partly explain why ice epochs are so rare on the scale of hundreds of millions of years of geological history.

But we have seen just how dependent we are—not just human civilization but all of life—on the Sun itself; and when we start to talk in terms of hundreds of millions of years we have to consider just how the Sun itself may change over long periods of time. It is reasonable to expect the Sun to rise tomorrow (and tomorrow, and tomorrow . . .) in much the same way that it has through recorded history, and to expect it to be just as bright tomorrow as it is today. But can we really say for certain that the Sun shone in the same way a hundred million, or a thousand million years ago? Even if the Sun is constant on average over thousands of millions of years (which is certainly what astronomers believe today), can we be sure that it never flickers for a cosmic instant? After all, a flicker lasting a million years would only be 0.1 percent of a thousand million years of steady burning, such a tiny fluctuation that it could be ignored as far as the Sun was concerned—but not as far as we are

*See *Our Changing Planet.*

41

concerned, with a million years representing an awesome span of time on the human scale. If the Sun sneezes, then the Earth, and human civilization, catches cold. So the time has come to see what we understand about the workings of stars in general, and of the Sun in particular, before coming back to look at the hazards of solar fluctuations.

## CHAPTER 4

# Birth of a Star

OUR SUN, DESPITE ITS OVERWHELMING SIGNIFICANCE AND importance for mankind, is just a star, by no means remarkable among the many stars in the sky except for the one rather important fact that it is so close to us. In round terms, the Sun is "only" 150 million kilometers away, and it forms the dominant central feature of the Solar System in which we live, with planets (including the Earth) and other cold, small bodies orbiting around its massive central fires. Today science has replaced religious mysticism in the study of the physical world—although perhaps the distinction is as much one of terminology as anything else, since modern scientists follow in a direct line from their inquisitive forebears, whether those earlier "scientists" were thought of at the time as priests, astrologers, or medicine men. What, then, can modern twentieth-century science tell us about that most important of stars (for us), the Sun?

The first mystery to be solved is that of the *existence* of the Sun, and although many other aspects of the Sun's present state were studied before any real insight into its origins was obtained, it makes sense now to look first at our modern understanding of how stars like the Sun have formed, and continue to be formed. The stars we see in the sky are all members of a great celestial family of stars, the Milky Way Galaxy, which forms a flattish disk some 100,000 light-years across and 2,000 light-years thick, containing about 100,000 million stars. This system forms a swirling mass with a spiral pattern rather like that of cream stirred into a cup of coffee, with bright stars and dark cold clouds of dust and gas mixed in the maelstrom. On the galactic scale of things, we are very far from being either alone or very special, and there is a multitude of other galaxies beyond our own Milky Way, scattered across the Universe.* But our central human interest in the Sun need take us no farther afield than the nature and structure of our own Galaxy to explain the existence of the Sun here today.

## BEFORE THE BIRTH

Stars live for many thousands of millions of years, and even their birth pangs take much longer than the life of an astronomer, or several generations of astronomers. So we can't get an idea of how the Sun formed by watching the entire birth of another star. However, in our Galaxy new stars are being created all the time, and we can see stars in all stages of development from birth (and before) to death (and after). From this variety around us, we can build up a picture of the life history of a star—just as we can get a good idea of the life history of a tree by going into a forest and looking at seeds, saplings, and mature trees (as well as dead wood), without sitting down patiently to watch one single tree growing. The story of how astronomers unravel their picture of the evolution of a typical star from studies of many separate stars, each at a separate state

---

*The story of the Universe beyond our Milky Way Galaxy is told in my book *White Holes.*

of evolution, is a fascinating one which has formed the theme of several books. But this is not the place to repeat that story. Rather, the main subject of this book is to set the scene for the latest studies of ways in which the Sun *differs* from the "normal." So, with apologies to all those dedicated astronomers whose work is being glossed over, I shall concentrate on the established picture of the birth of stars within our spiral Galaxy, without going into any detailed explanation of how the picture was obtained.

Young stars in the Galaxy seem to be associated with clouds of gas and dust between the stars, and this, astronomers are sure, is no coincidence. The interstellar clouds of dust and gas which line the spiral arms of the Galaxy seem to be the birthplaces of stars which are forming today.

This material between the stars is mainly hydrogen, which is the simplest atom of all and the basic building block of the material in the Universe, and dust grains, which may contain more complicated molecules built around a carbon base (more of these later). This interstellar matter forms vast clouds, which may contain hundreds of times as much material as there is in our Sun, but spread over a much bigger volume of space. By and large, these clouds are stable, and show no tendency either to break up into smaller clouds or to spread thinner and thinner until they dissolve into nothing. But sometimes the balancing forces which maintain these clouds can be tipped, so that they begin to contract, and break up as they collapse, forming an association of small, dense clouds. This is where the life of a star begins, for once a cloud becomes compact enough the pull of gravity inevitably squeezes it into a smaller and smaller ball, while at the same time the cloud heats up inside, as gravitational energy from the collapse is turned into heat energy. Once the embryonic star gets hot enough, conditions in its dense center become ripe for the start of nuclear fusion reactions in which hydrogen nuclei are fused together to make nuclei of

45

helium, the next most complicated nucleus, and liberating even more heat. This heat then builds up pressure inside the star and halts the collapse, counteracting the pull of gravity, for as long as the supply of nuclear "fuel" lasts. From the beginning of nuclear fusion, the life of the star proper begins, and will be discussed in detail in the next chapter. But how does a great cloud of interstellar material get as far as the nuclear burning state in the first place?

Just as it is no coincidence that young stars are associated with clouds of interstellar material, so it is no coincidence that both young stars and collapsing clouds seem to be associated with the spiral arms that are the most prominent feature of a galaxy like our own. Everything in the disk of the Galaxy is orbiting around the center of the Milky Way system, like a much bigger and more elaborate version of the way in which all the planets of our Solar System orbit around the Sun. But this rotation doesn't take place like a solid disk, such as a phonograph record. Rather, the bits and pieces closest to the center orbit fastest, while those in orbits farther from the center take a more leisurely turn about the nucleus. Our own Sun and Solar System, about two-thirds of the way out from the center of the Galaxy, take a couple of hundred million years to get round the system once. This may sound a long time by human standards, but still it means that the Sun has swung around the Galaxy twenty times or more in its known lifetime.

Before the Sun formed, there must have been a great interstellar cloud following much the same orbit around the galactic nucleus, until something gave that cloud a squeeze and started its collapse and breakup into stars. Now, the edges of the spiral "arms" are exactly the places where a cloud that is passing through will get a "squeeze" of this kind. Depending on your point of view, you can either imagine this spiral pattern sweeping around the Galaxy and past individual stars and clouds, like a wave passing across the sea and past individual molecules of water, or you can treat the wave as a stationary "standing shock wave" (analogous to the standing sound wave in an organ pipe) through which all the stars, solar systems, gas, and dust of the Galaxy must move as it orbits the center of the Galaxy. What matters is that the spiral feature does mark a region of compression in the swirling material of the Galaxy. If we take the natural viewpoint, from our position on a planet orbiting the Sun, that the compression zone is a stationary shock wave through which we pass in our travels, then the picture goes like this.

## THE GREAT SQUEEZE

As the material moving round the disk reaches the shock wave, it gets squeezed and compressed, building up as a dark lane like the flotsam swept along by an ocean wave. Once it is through the shock, the cloud of dark material can continue on its way, slightly more compact than before, until it meets the second spiral arm, with its attendant shock, on the other side of the Galaxy, where it is squeezed a little more. After a few of these regular encounters, with the cloud being squeezed smaller at each stage, the inevitable happens, with gravity taking over to pull the cloud down and break it up to form a clutch of new stars.

New stars are bright, burning fiercely with the exuberance of youth, and this explanation of star formation requires that right next door to the spiral shock wave there should always be a steady supply of new young stars lighting up as they move off "downstream" in their orbits. That is just what we see both in our own Galaxy and other spiral systems—dark lanes of compressed gas and dust, with bright spiral arms full of hot young stars next to them.

This picture is almost certainly correct, in broad outline at least, in explaining how vast, diffuse clouds of material get squeezed (over several orbits of the Galaxy) into a state where they are ready to begin the collapse into new stars. Astronomers are still puzzling over many of the details, however, and in particular there is a very strong hint that at least in some cases, including the formation of our own Solar System, an extra "trigger" may come in to set off the final collapse itself.

To understand this trigger we have to get ahead of ourselves a little, since it is related to the death, as well as the birth, of a star. As we shall see in detail in Chapter 5, one way in which a star can die is in a violent explosion, a "supernova" which spews a variety of material out into the interstellar medium, thus helping to regenerate the mixture of gas and dust between the stars and, in the long run, providing the raw materials from which later generations of stars can form. Among this material, supernovae scatter radioactive elements which can be distinguished by their characteristic radioactive "decay," in which these elements eject basic particles such as electrons and protons to become normal, stable elements. Because some of these elements can be produced only by the violence of an event like a supernova explosion, and because once formed they do not last forever but must decay into a stable element in due course, if they or their residues are found in the vicinity of the Solar System today that can only mean that a supernova exploded relatively nearby a relatively short time ago.

In this case, "relatively" recent means as long ago as 4,000 million years plus, around the time the Sun formed, and "relatively" nearby means where the proto-Sun was then, some twenty-plus orbits of the Galaxy ago. Not exactly just next door in the middle of last week! Fortunately, we carry the evidence of what things were like there and then about with us in the Solar System, in the form of meteorites. These little bits of rocky debris left over from the formation of the Solar System, which from time to time run into the Earth, have been analyzed by scientists. One such piece of cosmic debris, the Allende meteorite, a two-ton mass of rock that fell in northern Mexico in 1969, has provided the key evidence of a supernova link with the formation of the Solar System.

That meteorite contains just the kind of elements that ought to be produced by the effects of a supernova. Not, in fact, the original radioactive elements, after all this time, but the characteristic stable elements which are left over after those radioactive elements have decayed. The supernova must have occurred only a few million years before the formation of the solid rocks of the Allende meteorite—in other words, just at the time the presolar cloud was beginning to collapse into its clutch of stars. With such a cloud hovering on the edge of collapse, and a great explosion occurring nearby in space and sending its shock wave rippling across the immediate vicinity, squeezing and jolting everything in its path, we need look no further to find the final trigger on the steady progress of the presolar cloud from an anonymous, dispersed state into a collection of hot young stars, among them our Sun and its family of planets.

The story may not have ended there, either, since the radioactive decay of the unstable elements into stable isotopes itself releases heat, and according to one suggestion heat released by this process may have been responsible for the production of a molten core in our Moon, something which is very hard to explain in any other way, as the Moon is so small compared with a planet like the Earth. This is only circumstantial evidence, perhaps, but the evidence from various studies that the Moon must have once had a molten core does help to persuade astronomers that the supernova trigger for the formation of our Solar System really did happen.

Just as the whole Galaxy rotates, so the collapsing cloud which was to become our Solar System was spinning, so that it settled into a disk. The central regions, with most material and dominated by the hydrogen gas of the cloud, became our Sun, while the surrounding particles clumped together to form planets. Near the young Sun, light gases were blown away by the heat and the activity of the young star, leaving behind small, rocky planets—Mercury, Venus, Earth, and Mars. Further out, in the depths of the Solar System, the disrupting influence of the Sun's activity was less and planets were able to hang on to large quantities of lighter material, forming the gas giants—Jupiter, Saturn, Uranus, and Neptune. Between and around these planets other bits of debris became moons or asteroids, some occasionally, as in the case of the Allende meteorite, still being swept up by the planets proper. Even further out from the Sun are some more mysterious members of the Solar System, the comets, which occasionally send a representative diving down close to the Sun's heat, where it may become spectacularly visible for a short while before climbing away into the colder depths of space again. Just possibly the whole comet family may be a transient feature of the Solar System, hitchhikers picked up the last time the Sun dived through the dense material at the edge of a spiral arm, which happened very recently by astronomical standards. Other results of that later brush with the spiral shock wave could have much more profound effects for mankind, a theme to which I shall return in the central part of this book.

But no events in the subsequent development of the Sun would have any significance for mankind unless life had managed to get a grip on planet Earth early in the history of the Solar System. So it is a vital part of our story to check up on just how life as we know it did begin on our planet, eventually to lead to ourselves and the variety of life we see about us. It is even more appropriate to look at this intriguing aspect of the Solar System here since, it turns out, the origins of life, as well as the origins of stars, are now thought to lie in those mysterious dark clouds of matter in interstellar space.

## THE SEEDS OF LIFE

In astronomy today, any theory which lasts for a decade or so without being seriously challenged becomes the "classic" explanation of its kind. By these standards, the classic explanation for the origin of life has until very recently been the view that life formed, by and large, on the surface of the Earth and some time after the planet itself had formed. The theory holds that after the Earth formed, a mixture of chemical compounds stewing up in the atmosphere and oceans, energized by sunlight and the sparking flashes of lightning in thunderstorms, eventually built up complex organic molecules including those which could reproduce themselves—the precursors of life as we know it.

One possibility, from comparison with the giant planets which have retained their primeval atmosphere, and from other studies, is that the early Earth had large quantities of methane, ammonia, and other "reducing" compounds in its atmosphere, and from the 1950s onward various experiments have shown that if such a mixture is bottled up, along with water, in a sealed flask through which electric sparks or ultraviolet radiation are passed then, sure enough, more complex organic molecules are built up.

But there has always been a nagging doubt in some people's minds about whether even the 4,500 million year history of the Earth is long enough for this laborious process to have had time, starting from the methane/ammonia stew, to produce the abundance of complex life forms here today. And now, as a result of new observations made in the late 1960s and 1970s, an alternative theory of the real origins of life molecules has come forward to challenge the classic view.

51

Radio astronomers studying those cool clouds of interstellar material have found characteristic radiation at microwave frequencies which can only be explained by the presence in those clouds of complex molecules which are formed from many combinations of the basic atoms such as carbon (C), oxygen (O), hydrogen (H), nitrogen (N), and others. These include such exotica as formaldehyde ($CH_2O$), $H_2CNH$, HCCN, $H_3CCOH$, and others with still more complicated structures. So, while theorists still believe that a mixture of water, methane, and ammonia, activated by sunlight and lightning, *could* produce a buildup of life molecules, it is no longer necessary to assume that only such very simple molecules were present on Earth from the start. Molecules which are only one step away from amino acids —themselves the basic constituent of the "life molecule" DNA—are already present in clouds between the stars, the clouds from which new solar systems form. Surely the atmospheres of young planets must be laced with such molecules, life waiting in the wings ready to be called upon to play its part on the planetary stage from the very beginning.*

During 1977, a combination of new observations of even more complex molecules in space and a new theory of the origin of life combined to persuade many astronomers that a new classic model had arrived. The observations revealed the presence in interstellar clouds of cyanotriacetylene (H-C-C-C-C-C-C-C-N), the biggest organic chemical molecule yet discovered in space and only very slightly different from one of the amino acids which, in turn, produce proteins, nucleic acids, and genes. The theory, which came from one of the best known of living astronomers, Sir Fred Hoyle, working with his colleague Pro-

---

*For more about the story of the discovery of molecules in space, see my book *Our Changing Universe.*

fessor Chandra Wickramasinghe, explains how such complex systems could form in the harsh environment of interstellar space.

The story starts with carbon grains—soot particles—coated with ice and stuck together in tiny "grains" of interstellar matter. This is more than mere supposition, since grains of this kind have been found in meteorites, meteorites which themselves contain organic molecules. Such tiny dust grains provide an almost ideal place for the slow buildup of complex molecules. In a cloud of gas and dust grains in space, atoms must from time to time collide with a grain of dust, and such a carbon/ice grain provides an ideal surface to make the atom stick in place. So, when other atoms come along and collide with the same grain they have the opportunity to interact with one another, building up molecules. From time to time, collisions with other grains, or the impact of the energetic particles called cosmic rays, may knock molecules adrift to wander until they collide with another molecule-encrusted grain, allowing simple molecules to build up gradually into very complex systems.

All this takes time. But the beauty of it is that there is plenty of time for the random workings of repeated collisions to build up complex molecules; there are thousands of millions of years in the quiet depths of space before the end products get anywhere near a young planet. And, more controversially but certainly providing food for thought, Hoyle and Wickramasinghe take the theory further by arguing that conditions in space will produce a kind of natural selection, favoring molecules that are best able to stick to carbon grains and leading to a form of life itself. In their own words:

There seem to be two somewhat different types of selective process which could be operative under interstellar conditions. First, a competition for clump growth in the absence of disruptive agencies, for example, within a prestellar

53

cloud; and second, a competition for organic molecules (or even for sticking) in a more hostile environment in the presence of disruptive agencies. With the development of a cell wall, the last step could be to "split out" the inorganic grains which started off the whole process.*

So, according to this theory, we would be left with a little bag of organic material inside its own protective cell wall, with the original grain ejected and free to start the growth of another life seed. Such protected, self-contained bags of life could very well survive the hazards of collapse of the gas cloud and the formation of stars and planets, until a few of them at least made a new home in the primeval atmosphere of a young planet like the Earth. The theory also implies that all life on all planets starts out from the same basis, built up from the uniquely active atoms of carbon—carbon-based molecules are simply stickier than others. There is no reason to expect our Solar System to be anything out of the ordinary, and our two kinds of planet (small and rocky near the Sun, gas giants further out) are probably the only kinds around. Given those two choices, though, the best home for life of the kind that begins to evolve in interstellar clouds is a methane-rich planet like Jupiter, which leads to some interesting speculations well outside the scope of the present discussion.

What matters here and now is that the best evidence today suggests that the process which gives birth to a solar system *automatically* leads to conditions on the planets which are suitable for the development of life, at least in some cases. We see that the seeds of life are already there

---

*F. Hoyle and C. Wickramasinghe, "Prebiotic Molecules and Interstellar Grain Clumps," *Nature* 266 (1977): 241; see also their book *Lifecloud.*

from the beginning, and that the organic molecules which are the basis of life on Earth are found everywhere through our Galaxy (and even in other galaxies). The beginnings of life start to stir even before the birth of a star, and are a feature of the same interstellar clouds from which stars are born. From then on, of course, the evolution of life on the surface of a planet like the Earth is intimately linked with the evolution of the central star of the solar system—with the "life" of the star.

CHAPTER 5

# Life of a Star

THE BIRTH OF A STAR LIKE OUR SUN IS INTIMATELY CON-
nected with the nature of the spiral galaxy, the Milky Way, in
which we live. Clearly, though, some stars must have formed
first in order to create this structured galaxy, and the Sun is very
far from being an old star, by the standards of our Galaxy or
the whole Universe. Most probably, astronomers believe, the
Milky Way Galaxy formed from a huge cloud of original gas,
hydrogen and helium and virtually nothing else, which took on
its spiraling whirlpool shape as it collapsed and settled into a
disk. The earliest stars formed contained only hydrogen and
helium; some of these stars still exist and can be identified by
spectroscopic analysis of their light, which shows that their
atmospheres are, even today, deficient in heavier elements compared with younger stars. Others in the original stellar population of our Galaxy exploded, scattering the elements built up
by fusion in their interiors to provide the raw material for
second generation stars, and so on. The biggest stars burn up
their nuclear fuel most rapidly and explode most violently; the
process is repeated in succeeding generations. From spectroscopic analysis of starlight and the light from the Sun, astro-

57

nomers deduce that our Sun is at least a third generation star and maybe even further removed from the original stellar material of the Galaxy, because even its atmosphere contains discernible traces of quite heavy elements. And the presence of heavy elements in the Earth, in you and in me, is also evidence that the interstellar cloud from which our Solar System was made had already been pretty well "cooked" inside earlier generations of stars.

The Earth also tells us, indirectly, that it is indeed nuclear fusion that keeps the Sun hot. Geological evidence indicates that the Earth is some 4,500 million years old, and that the Sun has been shining for all that time. Ordinary chemical burning would long since have been exhausted—if the Sun were made of solid coal, for instance, it would have burned to ash long ago. And that heat released as gravitational energy is converted into thermal energy in a collapsing gas cloud is also totally inadequate to explain the heat output of the Sun over such a long time. Until the discovery of matter-energy conversion in nuclear reactions, astronomers were distinctly baffled by the Sun's longevity and prodigious output of energy; and it was only in the 1950s that astrophysicists really came up with a satisfactory detailed explanation of just which nuclear fusion reactions, going on inside the Sun, would serve to keep it on the boil and produce its present appearance today.

These reactions are the very earliest steps in the ladder toward production of heavy elements described in Chapter 2. Hydrogen is being converted into helium in the Sun's center at a rate of 600 million metric tons a second—yet the Sun contains so much hydrogen that this conversion can continue for about another 5,000 million years before the hydrogen "fuel" is exhausted. Since the geological evidence points to a present age of about 4,500 million years for the Solar System, the Sun today is almost exactly halfway through its life as a hydrogen burning star. During this phase of relatively stable existence, the Sun is

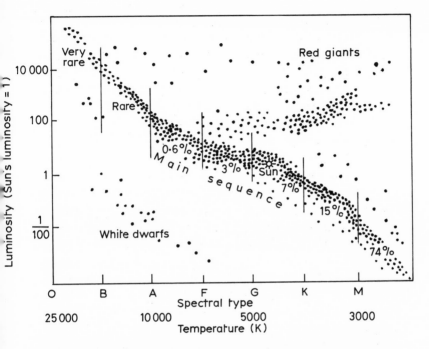

*Figure 4* The Hertzprung-Russell (H-R) diagram. When stars are described by their luminosity and surface temperature (or spectral type, which depends on temperature), they fall into three main categories: red giants, white dwarfs, and the main sequence. The Sun is a main-sequence star roughly in the middle of this plot; the percentages of each type of main-sequence star are marked for comparison—most are much fainter and cooler than the Sun. Here, as throughout this book (unless otherwise noted in the text), temperatures are given in Kelvin units (K); 0°C, the freezing point of water, corresponds to 273 K.

said to be a "main-sequence star," since all stars in this state have very similar properties, with their outward appearances differing only because they contain different amounts of material—they have different masses.

59

The main sequence is part of the "wood" which tells us how stars evolve without our needing to wait for thousands of millions of years patiently watching one star age. The whole "wood" is called the "Hertzprung-Russell diagram," after the two astronomers who developed the technique of labeling a star in terms of its brightness (compared to the brightness of the Sun) and the temperature at its surface. To find the brightness, astronomers need to know the distance to a star so that its true brightness can be worked out from its apparent brightness in the sky; to determine the surface temperature, they again use spectroscopic data, classifying stars in categories labeled (for historical reasons which no longer make much logical sense) O, B, A, F, G, K, M. In essence, the spectroscopic differences are those of color; just as a white-hot poker is hotter than a red-hot poker, so O stars are hotter than B stars, and so on down the scale. The Sun, on this classification, is a type G star with a surface temperature just under 6,000 K, corresponding to a yellowish color. O and B stars are blue, A and F white, K orange, and M red.

In fact, astronomers are able to make more subtle subdivisions and to assess the brightness and temperature of individual stars with great accuracy; the subtleties of the techniques, however, must be glossed over here as we concentrate again on what the findings tell us about our own Sun, rather than on the story of how the evidence was gathered in the first place.

When the temperatures of stars are plotted against their brightnesses on a Hertzprung-Russell (H–R) diagram, the resulting dots are not scattered about at random. Instead, there is a broad band running from top left (hot and bright) down to bottom right (cool and dim), the main sequence where most stars lie. In the bottom left of the diagram there is a scattering of hot but faint stars, called white dwarfs, and in the top right another scattering, this time of bright but relatively cool stars, the red giants. These names are logical and descriptive. A hot star can only be faint if it is very small, so that although the surface shines brightly there isn't much surface doing the shining, and a cool star can only be bright if it has a huge surface each square meter of which contributes a small amount of light to the total. So white dwarfs are small and hot, while red giants are large and cool.

The H–R diagram of any particular group of stars—say, all the stars in the immediate neighborhood of the Sun—tells you the state of those stars at one cosmic instant of time. As stars age and evolve, they move across the H–R diagram; but, as I have already mentioned, a star like the Sun may spend as long as 10,000 million years on the main sequence. When the Sun formed from a collapsing cloud of gas, it was certainly cool to start with, and large, and radiated a lot of energy. So it began life as a star above and to the right of the main sequence, in the red giant half of the diagram, although not as a red giant. Once the nuclear fusion in its interior became established and the Sun settled down, the temperature it ended up at depended only on its mass—the temperature in the middle of a star has to be exactly enough for the outward pressure to balance the inward pull of gravity, which depends on mass alone. The mass decides both the internal temperature (which in turn fixes the surface temperature) and the actual size of the star; so the mass fixes the place of a star on the main sequence.

A more massive star will be squeezed tighter by gravity, and must "burn" its nuclear fuel faster, getting hotter in the process, in order to remain stable. So the hottest, brightest main-sequence stars are the most massive (top left-hand corner of the H–R diagram), and also the shortest-lived, because their fuel is being burned quickly. Less massive stars need burn their hydrogen fuel only sparingly to be held up against the pull of gravity, so they are cool and dim (bottom right of the diagram), but long-lived. The bright, flashy giant stars burn out so quickly that few are around today—less than 1 percent of the stars in the Sun's vicinity belong to the upper main sequence, above spectral type F, where surface temperatures climb above 10,000° C. On the other hand, the quiet, small, and long-lived stars of category M, with temperatures of only about 3,000°C, make up almost 75 percent of the stellar population in our part of the Milky Way—and these figures are probably typical of other stellar neighborhoods as well.

The mass of our Sun is about $2 \times 10^{33}$ (2 followed by 33 zeros) grams; such numbers are rather unwieldy, and the Sun is far from being a large star, so to keep things simple astronomers generally say that the mass of the Sun is 1 (1 solar mass) and give the masses of other stars in terms of this basic unit. B stars have masses about sixteen times that of the Sun, while M stars have masses of only about one third of a solar mass. The effects of these differences on lifetimes are dramatic. While a star with, say, twenty times the Sun's mass has twenty times as much hydrogen fuel to burn, it will be 10,000 times brighter than the Sun and using that fuel up correspondingly quickly. So its lifetime on the main sequence will be only 20/10,000 times that of the Sun, one five hundredth of the Sun's lifetime, just 20 million years. A "star" with mass less than 5 percent of the Sun's mass, on the other hand, never gets hot enough in the middle for nuclear fusion to begin, and settles into a very compact, very dense lump once all of the heat liberated from gravitational energy by contraction has gone. In a sense, Jupiter can be regarded as a failed star of this kind rather than a true planet.

Our Sun, with its mass of $2 \times 10^{33}$ grams, has a main-sequence lifetime of about 10,000 million years; with only 25 percent more mass, this period of quiet stellar respectability would be reduced to a third of that, and a star of 1.5 solar masses would use up all its hydrogen fuel in less than 2,000 million years. This raises interesting questions for life, since here on Earth it has taken more than 4,000 million years of a reasonably well-behaved Sun to allow life to evolve to the point where intelligence has emerged. After the hydrogen burning stage on the main sequence, stars vary so much that inner planets, like the Earth, can be totally destroyed. It may be that once a star is much bigger than the Sun the main-sequence lifetime is simply too short for intelligence to have a chance to develop on any attendant planets. On the other hand, small, dim stars may simply be too cold to provide the vital energy on which planetary life depends; so stars much smaller than the Sun may also have no intelligent life on their planets.

This doesn't mean that there is no intelligent life elsewhere in the Universe, or even in our own Milky Way Galaxy. There are an awful lot of stars very much like our Sun, among the thousands of millions of stars in the Milky Way. But it may explain, rather neatly, why we live on a planet orbiting a G type star—these are quite possibly the best stars for life, warm enough to provide energy for organisms on their inner planets, and long-lived enough for those organisms to have a chance to develop intelligence.

In human terms, then, the "life of the Sun" is essentially its life *on the main sequence,* the quiet respectability of middle age, free from the outbursts of youth and the decay of old age, which provides a central furnace in the Solar System steady enough for life to evolve and intelligence to emerge. If it were not for its absolutely crucial importance to us, this 10,000 million years of steady nuclear burning, converting hydrogen to helium, might seem rather boring and dull. But knowing what makes the Sun tick today must have its fascination for all of us dependent on that steady ticking; and as astronomers now realize, a star that burns "steadily" on average for 10,000 million years can be subject to fluctuations which are small and rare by the standard of that quiet phase of its life, but large and long lasting by the standards of human civilization. We can only hope to understand these flickers in the solar furnace, and their implications for mankind, if we have at least a rough idea of what the "normal" state is, from which these departures occur.

The main change that is going on inside the Sun now is that the proportion of helium is building up in the core, and the size of the helium-rich core is steadily growing as the nuclear fusion region pushes outward from the center, where hydrogen fuel is depleted. Initially, the Sun was an even mixture throughout, 75 percent hydrogen and 25 percent helium (with an almost immeasurably small smattering of heavy elements). Today, some 5,000 million years later, in the heart of the Sun these proportions have been almost reversed, and the mixture is now approaching 65 percent helium with only just over 35 percent hydrogen; the outer two-thirds of the Sun, however, still have the original composition, and there is a smooth shift between the core and the outer layers, a transition zone in which the proportions of hydrogen and helium depend on the exact distance from the center of the Sun.

All this is deduced by indirect evidence—we cannot "see" inside the Sun and measure the hydrogen/helium ratio. Instead, astrophysicists make computer simulations—mathematical models—of stars, in which the equations relating mass, central pressure and temperature, density, and the equations of the nuclear fusion process are used to simulate mathematically the entire life history of a star from the start of nuclear burning to its end. The only specifications needed are the initial mass of the star and its initial mixture of hydrogen and helium; we know the mass of the Sun, and assume that the measured ratio of hydrogen to helium which we can see in the outer layers today remains the same as the original mixture. Then, when the equations in the computer are left to "run" for the equivalent of a stellar life of 5,000 million years, the end product is a star just like the Sun but with, the computer tells us, the internal composition described above. With a different internal composition, the same models predict that a star of one solar mass and 5,000

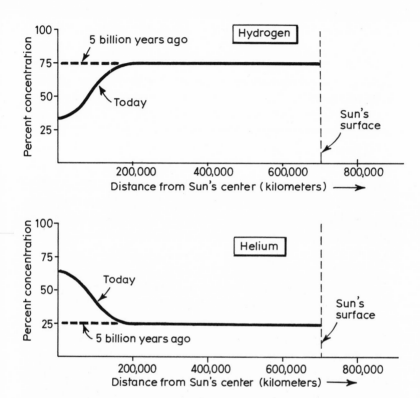

*Figure 5* The changing chemical composition of the Sun. Initially, 5,000 million years ago, the Sun was a uniform mixture of 75 percent hydrogen and 25 percent helium, indicated by the straight lines. Today, the ratio of elements in the core has been almost reversed as hydrogen fuel is "burned" to make helium—but plenty of hydrogen remains in the outer layers. On this scale, all the other elements contribute a negligible fraction to the Sun's chemical mixture.

million years old would not look like the Sun—it would be cooler or hotter, bigger or smaller. So unless astrophysics has made some very basic mistake, we know what is going on inside the Sun.

There are certainly no mistakes in the equations describing how pressure balances the structure of the star against the inward tug of gravity—those are very well known from studies here on Earth, and the physical laws behind them have been studied since the time of Newton. We are *almost* sure that we understand the laws governing nuclear fusion as well, but there is room for a tiny doubt. This may have important repercussions for life on Earth (see Chapter 7), but is not big enough to affect the overall picture. Even allowing for the small margin of possible error in our understanding of nuclear fusion, the broad picture is much the same.

And the broad picture is certainly impressive. Inside the inner one quarter of the Sun by radius (only 1.5 percent of the Sun's total volume), half its mass is concentrated and 99 percent of its energy is generated. The temperature is about 15 million degrees at the center, a tiny bit higher just outside the center where the main nuclear fusion region is today, and then it falls off dramatically down to the surface which is about 6,000°C. The density of the core is 160,000 kilograms per cubic meter—twelve times the density of lead—but falls so rapidly that halfway to the surface the density is only the same as that of water. Because of the great central density, however, by halfway to the surface 90 percent of the Sun's total mass has been included; the outer half of the Sun contains only 10 percent of its mass.

It is these extreme conditions that make the center ripe for nuclear fusion. Under such extremes, our everyday ideas about matter being solid, liquid, or gas become irrelevant, and we have to think in terms of a fourth state of matter, plasma. In this state, electrons are stripped from their atoms to produce a fluid sea of positively charged atomic nuclei, swimming through a sea of electrons and occasionally colliding with one another. Those collisions are occasionally hard enough to encourage the nuclei to fuse together, releasing energy, keeping the Sun warm and building up heavier elements in the process.

The energy produced in the interior is in the form of electromagnetic radiation, X rays, and the like. And because it is produced in a plasma, it can escape very easily—up to a point. Electromagnetic radiation interacts very strongly with electrons that are tied to nuclei, in other words, with atoms. But it interacts much less easily with either free electrons or free nuclei. Whereas electrons tied to atoms can *absorb* radiation, "loose" electrons can only deflect radiation, bouncing it around inside the plasma. So the radiation from the Sun's interior proceeds blithely on its way, at the speed of light, bouncing around and working its way outward to the point where the temperature inside the Sun is low enough for atoms to exist. This covers the inner 85 percent of the Sun's radius. (The total radius of the Sun is about 109 times that of the Earth, some 1,390,000 km.) At this point, the radiant energy escaping from the Sun's core is blocked off as if by a black curtain; the atoms (actually partly ionized) in the layer immediately above the plasma region absorb all the energy, and get hot in the process.

The result is convection. Just as hot air rises (assuming that there is colder air above it), so the hot material about 200,000 km below the surface of the Sun rises outward in great blobs, displacing cooler material above which sinks into the heated region, warms up itself, and rises in its turn. The effect is a bit like those decorative lamps which contain a thick, colored oil immersed in a clear, lighter fluid and warmed from below by the heat of a light bulb. The hot material rises, gives up its heat to cooler layers above, and falls back, in a pattern of convective circulation. Exactly the same kind of convection currents develop in a pan of water (or porridge) heated on a stove; the main difference is that on the Sun the source of energy is radiation

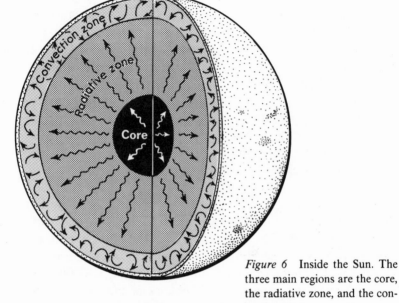

*Figure 6* Inside the Sun. The three main regions are the core, the radiative zone, and the convection zone.

from the nuclear furnace of the core, and, of course, the scale is rather more dramatic.

Astronomers believe that the convective zone, the top 150,-000 km of the Sun, is arranged in three main layers, with huge circulating currents perhaps 200,000 km across just above the radiative interior, smaller granulations about 30,000 km in diameter above them, and small convective loops 1,000 km across and 2,000 km deep, forming the very top layer, the visible surface of the Sun. At this point, although the material is still cool enough for electrons to be attached to atoms and able to trap electromagnetic radiation, the solar material has thinned out so much that there are few atoms left to do the trapping— we have reached the surface of the Sun, and once again radiation can escape, this time out into space as sunlight. The intense

68

radiation of the nuclear processes in the core has been through many changes in its struggle out to the surface. Although, as I have mentioned, radiation cannot be *trapped* by the material in the radiative zone, and although the radiation does travel at the speed of light (30,000 million centimeters a second), it does not travel outward in a straight line. The particles in the plasma, both electrons and nuclei, but chiefly the electrons, can *scatter* electromagnetic radiation, bouncing it around on such a tortuous path that it takes a million years for energy to struggle from the center to the outer layers of the Sun. There, carried outward by convection for the last 15 percent of its journey, the energy moves much more rapidly and directly, before being converted back into radiation and escaping into space, where it can move straight outward at the speed of light.

One result is that while it takes a million years for energy to get from the inside of the Sun to the surface, a journey of 1,390,000 km, it takes only 500 seconds (just over eight minutes) to travel the further 150 million km to the Earth. But there are much more important consequences. First, the slow spread of energy out through the Sun helps to keep the interior as a whole steady; everything emerging from the deep interior is averaged over a million years or so by the time any influence is felt by the convective zone and the surface. So any small fluctuations deep inside the Sun which last for less than a million years literally have no effect on the Sun's outward appearance. If, by magic, we could switch off the nuclear burning for a day and then switch it back on, the Sun would *not* "go out" like a light, and no discernible effects would be felt by life on Earth.

That is certainly comforting, although it is difficult to see how the Sun could switch off for a while (or is it? more of this later!); but there is another side to the coin. If the surface layers are indifferent to conditions inside the Sun today, but depend only on the average conditions inside the core over the past million years or so, then *we cannot make any direct inference about exactly what is going on inside the Sun now by looking at the surface layers.* And the surface is all we can look at with optical telescopes, radio telescopes, X-ray instruments onboard satellites, or anything dependent on electromagnetic radiation.

So whenever we talk about what the Sun is doing now, how it got to the state it is in today, and where it is going in the future, we depend upon a combination of the computer models mentioned above and observations of many stars like the Sun, some older and some younger, from which we infer, with the aid of relationships like those expressed in the Hertzprung-Russell diagram, what the *typical* behavior of a star like our Sun is, and how it will evolve *on average* over the next thousands of millions of years. Any subtleties—temporary departures from the "normal" pattern, or permanent slight differences between the Sun and other stars—have to be worked out by other means.

Even so, astrophysicists can describe with confidence the broad pattern of the rest of the Sun's life, both on the main sequence and after.

During its life on the main sequence, the Sun expands slightly in size as the core shifts and adjusts to the changing mixture of hydrogen and helium it contains; but this small effect has no great significance for life on Earth. But when all of the hydrogen in the core—perhaps 10 percent of the original mass of the star—has been converted into helium, much more dramatic changes take place. This happens when our Sun is about 10,000 million years old (in about 5,000 million years time), as it moves into old age. With no more hydrogen to burn in the core, no more heat is generated at the center of the star by nuclear fusion, so it is no longer held up against the pull of gravity. Inevitably, the core begins to collapse, and as it does so gravitational energy is converted into heat, just as it was when the young star first formed from a collapsing gas cloud.

This has a curious effect. The extra heat from the center causes the outer layers of the star to expand dramatically, even while the core is still collapsing. The star becomes bigger and brighter, moving upward on the Hertzprung-Russell diagram, but the surface becomes cooler, shifting the star to the right on the diagram. The Sun will become a red giant, taking a few hundred million years to swell up to 100 times its present diameter, drastically affecting the ecological balance of the Earth and probably destroying all life on our planet.

During this shift into the red giant state, the region just outside the collapsing central core becomes hot enough for nuclear hydrogen burning to take place in a shell around the core, where hydrogen was never used in the main-sequence, core-burning stage of life. At the same time, the core itself contracts further and gets still hotter, until at about 100 million degrees conditions are right for the next stage of nuclear fusion, and helium burning begins, converting helium nuclei into carbon and oxygen nuclei. Now the star has two sources of energy, the core stabilizes, and the star turns back to the left on the Hertzprung-Russell diagram, toward the main sequence. For a star like the Sun, the switch-on of helium burning happens very suddenly, according to the computer models, in a burst dubbed the "helium flash"; for stars more than twice as massive as the Sun, the core helium burning begins in more leisurely fashion at the very center. Either way, with all these changes inside the star and constant readjustments of the outer layers in response, life as a red giant is more exciting than life on the main sequence, with stars following wiggly paths to and fro across the red giant part of the Hertzprung-Russell diagram as they age.

Their ultimate fates depend, once again, on their masses. In the case of the Sun, the helium flash is brief but dramatic, with readjustment taking only hours and being followed by a phase of giant growth lasting perhaps 30 million years. During this time the Sun, swelling up to 400 times its present size, will

71

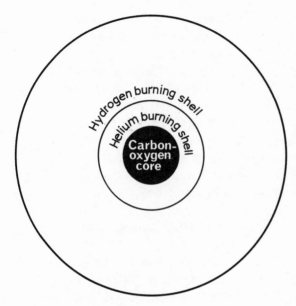

Hydrogen burning shell

Helium burning shell

Carbon-oxygen core

*Figure 7* The Sun in old age. The heaviest elements a star like the Sun can build up—late in its lifetime—are carbon and oxygen. As the Sun ages, an inert core of these two elements will be surrounded by a shell in which helium is being "burned" to make more carbon and oxygen, and by a shell in which hydrogen nuclei are being fused into helium nuclei. These shells move outward through the star with the core growing bigger, until the nuclear fuel is exhausted and/or the outer layers are lost into space.

swallow all the inner planets, Mercury, Venus, Earth, and Mars, while losing material from its bloated surface into interstellar space, matter which will help to form the next generation of stars. But when the helium fuel at the center is exhausted, the Sun will have no reserves left to draw on. A star bigger than about 4 solar masses has enough potential gravitational energy to warm its core, by collapse, to a stage where carbon burning begins, to be followed in its turn by further steps up the fusion ladder at hotter and hotter temperatures until something has to give and the star explodes—more of this later. But the Sun will never get hot enough to burn carbon; after a phase of helium shell burning around a collapsing core of carbon/oxygen "ash," there are no more reserves of nuclear energy that the Sun can draw on, and the end of its life will be in sight.

72

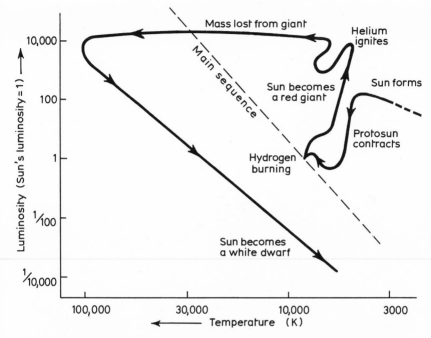

*Figure 8* The Sun from birth to death. The life history of a star like the Sun is shown here as a path across a stylized H-R diagram (see Figure 4). The time spent in each part of the path is roughly 10 billion years of hydrogen burning on the main sequence, up to 2 billion years as a red giant, and then a rapid contraction into a burnt-out white dwarf state, which will last forever. (Based on Figure 7.7 of W. J. Kaufmann's *Stars and Nebulas.*)

With no more fuel for the nuclear furnace, gravity will at last win its long-drawn-out battle and pull all of the Sun's remaining material into a small, cooling lump. Of course, the last collapse itself will heat the stellar cinder up, so that it becomes small and hot, shifting from the red giant branch of the Hertzprung-Russell diagram right across to the lower left-hand side, becoming a white dwarf. As the star cools, it becomes first a red dwarf and then, ultimately, a black dwarf emitting no light at all and chilled to the cold of space itself. With a density inside of 1,000 metric tons per cubic inch, the star is held up by a different kind of pressure, the pressure resulting from the fact that there is simply a limit to the number of electrons that can be squeezed into a small volume—you can't get a quart of electrons into a pint pot, so the star stops shrinking at the quart pot stage.

For the Sun, the "quart pot" is about the size of the Earth —so that is the size of the cooling white dwarf, surrounded by a dispersing cloud of blown-away material, that will mark the site of the once proud Solar System in about 6,000 or 7,000 million years' time. But this is only of academic interest to us. The ultimate death of the Sun is a pretty boring affair, nowhere near as dramatic as the fate of the short-lived, massive stars that go out not with a whimper but a bang, and form the subject of the next chapter. But armed with this knowledge of how stars are born, live, and die, we will be ready to return to the puzzles of just how steady the Sun's main sequence lifetime really is, and how an insignificant flicker for the Sun can shake the whole Earth and all life on it.

CHAPTER 6

# Death of a Star

THE DEATH OF THE SUN ITSELF, AND THE DEATH THROES which will engulf everything in the Solar System even beyond the orbit of our planet Earth, is dramatic enough by any human standards. But, in fact, it is pretty small beer by astronomical standards, not just because there are so many stars around that one more or one less is hardly worth noticing, but because some stars end their lives in such spectacular fashion that the death of the Sun, by comparison, will be like a flickering candle alongside a nuclear bomb. Long life on the main sequence, quietly burning nuclear fuel at a modest rate, goes hand in hand with an unspectacular demise in stellar terms. Ninety percent of all the stars in our Galaxy have about the mass of our Sun or less, and virtually all of these small but common stars are still living their lives, as described in the previous chapter. The remaining 10 percent, though, lead shorter, sharper, and more spectacular lives before going out in a blaze of glory; although only 10 percent of the total *number* of stars in the Milky Way, they make up 25 percent of the *mass* of the Galaxy, and some are many times as massive as our Sun. The bigger they are, the quicker they burn their basic nuclear fuel in the effort to balance the tug of gravity, and the more spectacular is their end when it comes—in their case, sooner rather than later.

All of these massive, short-lived stars are important to us. It is in their nuclear furnaces that the heavy elements of which we are made are produced, and it is in their explosive death throes that the heavy elements are scattered to provide the building materials for planets like the Earth and stars like the Sun. They also produce some of the most interesting objects in our Galaxy, compact remnants of once great stars, in the form of tiny but energetic pulsars, and perhaps even black holes. Here, both space and time become distorted in barely conceivable ways, and even our very best theory of the nature of the Universe, Einstein's General Theory of Relativity, breaks down and becomes inadequate.

## THE NOVAE — HALFWAY TO GLORY

Between the common, quiet stars like the Sun and the biggest and brightest stars there is a range of masses where life and death follow the same pattern as for a star like the Sun, but with everything painted with a bigger brush. Evolving rapidly as the basic nuclear fuel in their centers is exhausted, in only a few million years such stars become red giants; but then, instead of a relatively quiet decline down into the white dwarf state, with just a few flickers along the way, they can release enough gravitational energy through collapse to start further cycles of nuclear burning, producing carbon in their cores. As this happens, the star becomes hot and blue, and changes in the nuclear burning can cause regular rhythms to pulse through the whole star, so that it swells a little then shrinks, as constant fine adjustments to the balance between gravity and pressure occur in its middle. These regularly pulsating stars are called Cepheid Variables, and their variations follow very precise periodic fluctuations, from a few days to fifty days long, depending only on their average brightness (that is, on their mass).

After this stage, as the core becomes dominated by carbon, the inner layers of the star shrink in upon themselves while the outer layers grow into a vast tenuous atmosphere from which material streams away into space. The star is, once again, a red giant; but it still has the capacity to go through the whole cycle again and again, building up the material in the core through successive phases of pulsation, collapse, and giant-hood until the core is chiefly composed of iron, the most stable element of all. Then, all bets are off.

No more nuclear fusion processes can release energy in the core, so that it must cool and collapse. With no heat from the inside coming out, there is nothing to hold up the outer layers

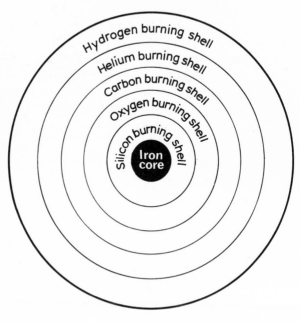

*Figure 9* An old massive star. With enough matter to play with, a star much more massive than our Sun can build up heavier elements in a succession of layers (shells) around a growing core rich in iron-56. Eventually, an explosion will disrupt such a star as instabilities develop, and in the process even heavier elements are formed with the whole stew being spread across interstellar space to provide the basis for later generations of stars and planets.

77

anymore, and they too collapse toward the center of the star, releasing gravitational energy and getting hot in the process. But these outer layers contain plenty of material like carbon, which can still form heavier elements by fusion when the temperature rises, so that in the middle of the collapse the suddenly heated outer layers literally explode, with a lot of carbon and other elements fused into heavier elements at once. The explosion blasts the outer layers away from the star, dispersing perhaps 1 percent of the total stellar mass into space.

These stellar explosions are visible for a long way across the Galaxy, because the energy released makes the star suddenly shine brighter than ever before. This makes astronomers notice them, and they are called novae (which means "new" stars) because they may suddenly appear brightly where no star has been noticed before. But they are not really "new," just a sudden flare-up of a once insignificant star. After these outbursts, novae have had their fling, and the material left behind can now settle down into a solid lump, free from nuclear fusion, held up simply by its structural strength and cooling off over millions of years to become a black dwarf, a cold lump of iron, perhaps covered with a skin of carbon or frozen carbon dioxide, of no interest except as a possible hazard to any space-faring race that may have discovered the secret of interstellar travel.

Novae are bright and dramatic, and fairly common—forty occur each year in our Galaxy. But they just miss the really spectacular form of star-death, the glorious end that is the exclusive fate of the very massive stars.

## SUPERNOVAE

If the choice of the name "novae" for what are really old, dying stars is something of a misnomer, the name "supernovae" for the most spectacular star-deaths is even more bizarre. The name is arrived at by taking the most obvious feature of a nova, its brightness, and combining this with the most obvious feature of a supernova, which is that it is many thousands of times brighter still. It is, in fact, "superbright," and neither new nor very new; but the name has stuck, and supernovae they remain.

These are stars with eight or more times the mass of our Sun, which go through the various processes of nuclear fusion right up to stable iron, producing a core rich in iron which is itself as massive as our Sun, or even more massive. With seven or more solar masses of material still pressing down on this core from above, and no more energy available from fusion to hold up the star against gravity, very interesting things begin to happen in the middle of such a star.

First, although there is no more stable element than iron, there are even more efficient ways of arranging the elementary particles from which atoms are made than as a sea of iron nuclei surrounded by electrons. The core becomes so dense that electrons (negatively charged elementary particles) and protons (positively charged nuclear particles) are squeezed together, combining to form neutrons, the uncharged counterparts of protons. The resulting sea of neutrons needs much less space than two intermingling seas of electrons and protons, so very suddenly the center of the star collapses dramatically. For the upper layers, it is as if the bottom dropped out of their world, and they fall inward into the gap created, releasing heat and initiating a burst of fusion similar to that of a nova but on a vastly greater scale.

With gravity now playing the dominant role for the first time in the star's life, iron nuclei in the core are ripped apart and turned back into helium nuclei. As much gravitational energy is put back into the core, from the collapse of the outer layers, as had been stored up by nuclear fusion throughout the entire previous history of the star. Still the energy left over is sufficient to tear the star apart as an energetic shock wave blasts outward

79

from the inner regions, and for a brief period this one star will shine as bright as all the stars of the Milky Way put together, pouring out in a matter of weeks as much energy as it had radiated throughout all the millions of years of its existence up to that point.

The debris produced in such an explosion spreads far and wide into the interstellar medium to contribute to the material of later generations of stars; and it is in these unimaginably huge explosions that the energy is available to make a smattering of very heavy elements, heavier than iron, including the radioactive elements like uranium and plutonium from which we can recapture a little of that energy today. Our fission reactors are powered by the energy of past supernova explosions, locked up in these heavy, radioactive elements.

Remnants of relatively recent supernovae can be seen in our Galaxy, as spreading clouds or veils of material. The most spectacular example is the Crab Nebula, the remnant of a supernova observed by Chinese astronomers in A.D. 1054; supernovae occur once or twice every hundred years in a Galaxy like our Milky Way, and astronomers reckon that by this rule of thumb we are overdue to see one relatively nearby, since only two others (in 1572 and 1604) have been seen by astronomers on Earth since 1054. But we can still study supernovae, because they are so bright that these explosions of individual stars can be picked out even in other galaxies, beyond our Milky Way, where one star may suddenly shine as bright as all of the thousands of millions of other stars in that galaxy put together.

But we don't want the next visible supernova in our Galaxy to be *too* close. As well as visible light, such explosions produce intense radiation (X rays and gamma rays) and floods of particles which travel at nearly the speed of light, the so-called cosmic rays. A supernova less than about thirty light-years away could produce dramatic effects on the Earth, especially as a result of these particles ripping into the upper atmosphere and

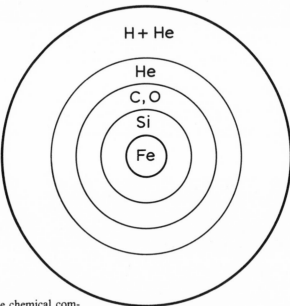

*Figure 10* The chemical composition of a highly evolved massive star.

doing severe harm to the ozone layer of the stratosphere, which protects life on the Earth's surface from the burning ultraviolet rays of the Sun. On the law of averages, such a damaging nearby supernova can be expected every 200 million years or so, so there isn't much chance that we will suffer one in our lifetimes. But, equally, the Earth should have suffered in this way several times during its long history, with damaging consequences for life. So some astronomers and paleontologists have suggested that the occasional rare but dramatic setbacks in the evolutionary record, when dozens of species seem to have been wiped out in an eyeblink of geological time, may be connected with nearby supernova explosions. In particular, a very dramatic change in the kind of creatures living on the Earth occurred around 65 million years ago, when the age of the dinosaurs came to an

81

abrupt end and was followed by the rise of the mammals, including ourselves. The suspicion is that the dinosaurs were wiped out by the stresses of a nearby supernova which disrupted the Earth's atmosphere and made their old way of life impossible, while smaller mammal-ancestors survived the holocaust and took over the ecological niches the dinosaurs left untenanted. It may be that we owe our existence to a supernova explosion, and we could just as easily see mankind's civilization destroyed by another such event. The chance of this happening is remote—far less than one in a million—but it is nonetheless real, and provides a chilling reminder of the insignificance of mankind on the cosmic scale of things.

## THE AFTERMATH

We don't know if any young civilizations were wiped out by the spreading blast from the Crab supernova, but we do know what the explosion left behind. In a white dwarf star, the matter is fantastically compressed, so that the mass of the Sun may be contained in a sphere the size of the Earth. But it is still understandably matter as we know it on Earth, and there is a meaningful way in which we can describe different layers as being made up of the nuclei of iron, or carbon, or helium. But when electrons and protons are squeezed together to form neutrons, as happens in the core collapse preceding a supernova explosion, all of these conventional labels become meaningless. Once nuclear fusion has run its course, any star with a mass of more than about 1.5 solar masses has a strong enough gravitational pull to do this trick, and the result is one huge "droplet," perhaps twenty miles in diameter, composed of nothing but neutrons. The only analogy we can make is with the nucleus of an atom—it is as if a whole star had been turned into one atomic nucleus twenty *miles* across, with such a ludicrously high density that if there were any way of measuring it one cubic inch of material from the star would weigh about 10 billion (million million) metric tons! This is a neutron star.

The neutron itself was only discovered by atomic physicists in 1932, and a year later astronomers suggested that neutron stars might form the ultimate final state of stars more than one and a half times as massive as the Sun. But it seemed there could be no way to study such an object, as it would be cold, with no internal heat left, and therefore invisible to ordinary telescopes. It seemed like an entertaining idea for the theorists to play with and nothing more. In the late 1960s, however, radio astronomers at the University of Cambridge made a discovery which really ought to have been predicted by the theorists, but somehow never was.

What they found was a completely new kind of radio source in our Galaxy: several objects in different parts of the sky which produce regular "pulses" of radio energy with periods close to one second, almost exactly like signals from some kind of beacon or radio "lighthouse." At first the Cambridge team thought they might have found just that—signals from some other civilization in space. But as more and more of these sources, soon dubbed "pulsars," were discovered, it became clear that they must be some natural phenomenon. The discovery of a pulsar in the Crab Nebula, pulsing once every 0.03309 seconds (thirty times in one second) clinched the matter, and revealed to the theorists just what the radio astronomers had discovered.

Any object which varies so rapidly must be very small, by astronomical standards. A large star "breathes" in and out over a matter of days (like the Cepheid variables) and only something as small as a white dwarf or a neutron star could vary with a period of a second or so without being broken apart by the forces involved. It doesn't matter whether the variation is due to literally expanding and contracting, or whether it is rotating and the radio noise comes from one spot on the spinning surface. To change that quickly the star has to be small.

The Crab pulsar clinched the choice between neutron stars and white dwarfs, because even a star as "big" as a white dwarf (the size of the Earth) cannot vary thirty times a second without falling apart under the strain. The Crab pulsar had to be a neutron star—and by implication so did all the others, since they all had much the same properties. In addition, the pulsar was found in the middle of a supernova remnant, exactly where we expect to find neutron stars!

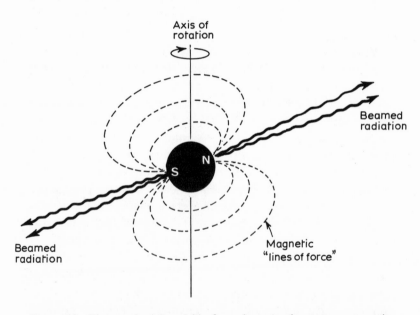

*Figure 11* The standard "model" of a pulsar. As the neutron star spins rapidly, with its magnetic axis tilted off the spin axis, particles and radiation escape from the polar regions. Each time a beam of radiation flicks across the line of sight we "see" (at radio frequencies) a blip of noise. Exactly like a lighthouse, the blips repeat at very precise, regular intervals as the star rotates.

What makes a neutron star pulse? This is where the theorists were kicking themselves for not predicting the existence of pulsars. All stars have magnetic fields, produced by the dynamo effect of moving charged particules in their interiors. These magnetic fields are locked into the structure of the star, so that if the *matter* of a star collapses the magnetic field is squeezed down with it into a smaller volume. The effect of squeezing a magnetic field is to make it stronger, and neutron stars possess the squeezed-down "fossil" remains of the magnetism of their parent stars. In addition, since all stars rotate (the Sun does so every twenty-seven days), neutron stars must spin. Like an ice skater pulling his arms inward, the star spins faster the more it contracts, so that a neutron star is not, after all, a quiet droplet sitting blankly in space, but must be a fast-spinning object, wrapped in a strong magnetic field and surrounded by electrically charged particles left over from the supernova in which it was born.

This is an ideal recipe for producing radio noise. Charged particles and strong magnetic fields produce radiation at radio frequencies, and the spin of the pulsar sweeps the radiation, leaking out from the magnetic poles, round in a circle. Each pulsar will produce two beams of radiation, and we "see" one of these with our radio telescopes when it sweeps across the Earth—it is exactly like a radio lighthouse, but one produced by nature rather than by any intelligent supercivilization. And the theorists could have predicted all of this, including the speed of the pulses and the likely locations of pulsars in supernova remnants, had anyone thought to ask the right questions in the early 1960s!

We can also explain why some pulsars pulse more rapidly than others—they are the youngest (like the one in the Crab Nebula), still spinning at the rate they were left with after the supernova explosion. Older pulsars slow down as they lose energy (through the beamed radiation), and they also get weaker; more than 100 pulsars are now known, but only a couple have enough energy to produce visible flashes of light, as well as radio noise, including the one in the Crab.

The discovery and explanation of pulsars was a key feature in revolutionizing astronomy as the 1960s gave way to the 1970s. For a start, the most way-out predictions of the theorists, ideas previously thought of more as science fiction than as respectable science, were seen to apply to the real Universe. And, secondly, the pulsars showed our Galaxy to be a place of violence and change, overturning the old picture of quiet stability. The advent of X-ray astronomy, dependent on detectors carried above the absorbing layers of the Earth's atmosphere in satellites, took things a stage further, revealing a breathtaking panorama of violent activity across the Milky Way, and providing confirmation of the reality of a concept even more fantastic than neutron stars—black holes.

## BLACK HOLES

If the matter inside a star more massive than one and a half Suns is squeezed so much that electrons and protons blend into one another to make neutrons, what happens at even more fantastic densities? Those same theories that predict the presence of neutron stars tell us that above 3 solar masses even the neutrons get squeezed—not so much into one another, but literally out of existence as the whole star collapses into nothing. Gravity wins its last battle, and all of the matter in the former star disappears into a mathematical point, a singularity.

This is where the theory breaks down—even General Relativity can't describe what happens at a singularity, and all we can really say is that the laws of physics as we know them no longer apply. Maybe the matter squeezed into a singularity in one place emerges somewhere else; since both space and time as we know them are also squeezed in such an ultimate collapse, it is even possible that it emerges some*when* else.* But as far as we are concerned here and now, what matters is the appearance of this collapsed star (or "collapsar") from the outside. We can leave aside these metaphysical mysteries, because there is no way we can ever see the singularity itself—nature applies a law of cosmic censorship, cloaking the singularity in an impenetrable blanket, called the event horizon, from which nothing can emerge. It becomes a black hole.

The story of black holes has been the stuff of several books (the best I know is *The Cosmic Frontiers of General Relativity,* by William Kaufmann), and I don't plan to go into the exotic details here. In a nutshell, though, what happens is that the intense gravity of the collapsar is so strong that nothing can escape its grip, not even light, which is why the "hole" is black —or rather, why it would be black if nothing fell into it. A collapsar sitting quietly on its own in space would be invisible and virtually undetectable. But one that is being fed extra bits of material from outside, new matter dropping down the gravity well of the black hole and gaining heat energy from the conversion of gravitational potential energy in the process, soon gets surrounded by a glowing cloud of material radiating across the electromagnetic spectrum.

For all practical purposes, the event horizon—the surface from which nothing can escape—around the singularity marks the surface of the black hole. Yet while a neutron star has more than a solar mass compressed into the size of a large island like Manhattan, the event horizon of a singularity that has swallowed several solar masses encompasses a region no bigger across than a small mountain, or perhaps Central Park in New York. The gravitational pull of several solar masses sucks matter into this tiny volume, and it is hardly surprising that in the

*See *White Holes* and *Timewarps.*

87

process the matter can pile up in a very congested "throat" leading down into the hole. The infalling matter piles up, the atoms in it collide violently with one another, and gravitational energy is converted into heat. Soon, the region just outside the black hole proper is a seething cauldron of reactions, perhaps complicated by the presence of magnetic fields, with radiation climbing across the spectrum from radio noise, across the visible bands, and into the ultraviolet, gamma ray, and X-ray frequencies. A black hole on its own is just about the ultimate in undetectability. A black hole provided with a source of extra matter, on the other hand, is one of the most spectacularly visible objects around. And very many black holes do indeed have handy sources of material nearby, in the form of other stars.

## BINARY SYSTEMS

A lot of stars—perhaps the majority of all stars, though we can't be sure—don't occur singly, like our Sun, but in pairs or more complicated groups, which formed together out of the same original collapsing cloud of interstellar material. Some of these pairs are widely separated, orbiting each other at distances comparable to the distance between the Sun and Jupiter, or greater, and evolve without having much influence on one another. Others are close binaries, and spend their lives in a complicated dance, exchanging matter as first one, then the other, swells to the red giant stage and sheds its atmosphere. This complicates their evolutionary patterns considerably. The star that evolves more quickly—the more massive star—expands first and dumps material on its companion, which might then become the more massive of the pair as a result, speeding up its evolution until it dumps matter back on its companion, and so on. The havoc this plays with the delicate balance between pressure and gravity makes stars in binary systems particularly prone to violent ends (some theorists believe that *only* stars in close binaries explode as supernovae). As a result, there are a lot of leftover neutron stars, and presumably black holes, orbiting around rather disturbed companions which are still trying to sort out their evolution at the red giant stage before settling down as white dwarfs or neutron stars themselves.

This provides a nice source of material from the outer layers of the confused red giant, spilling onto the tiny companion. Even if the companion is a neutron star, the energy released as the matter falls through its magnetic field and onto the surface is enough to generate X-radiation, and sure enough X-ray as-

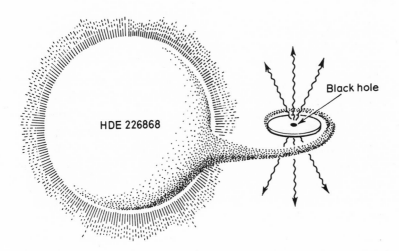

*Figure 12* The standard "model" of a black hole in a binary system, with Cygnus X-1 (HDE 226868) as the archetype. Material from the large star streams out as a stellar wind to be trapped and pulled into the black hole. Falling gases are squeezed and heated as they fall down the gravity well of the hole, releasing intense bursts of radiation at X-ray frequencies.

tronomy has now revealed the presence of X-ray pulsars in binary systems. "Ordinary" pulsars may result when the supernova explosion of one star disrupts the binary pair, sending the young neutron star scooting off across space on its own. But at least one of the X-ray sources now known may very well be a black hole orbiting around a relatively ordinary star.

The candidate is a source called Cygnus X-1, identified with a radio source called HDE 226868 and a hot, massive blue star in the constellation of the Swan (Cygnus). The star itself is a member of a binary system, and the radio and X-ray emission comes from its companion, which is invisible to ordinary telescopes because it radiates very little energy at optical frequencies. By monitoring the way in which the star associated with Cygnus X-1 moves around its companion, and having a rough idea of how massive such a hot, blue star must be from their general understanding of stellar evolution, astronomers can make a good estimate of the mass of this invisible companion that is such a powerful source of X rays, and the answer they come up with is at least 8 solar masses. There is no way such a massive object could be a star on the main sequence and not be visible; it could not be a white dwarf or neutron star accreting matter from the larger star, since the mass greatly exceeds the black hole limit; so it must be a collapsar, surrounded by a cloud of hot material producing the X-radiation, 10,000 times as much energy as X rays, as our Sun emits as ordinary light!

## THE VIOLENT UNIVERSE

So today the ultimate prediction of theory, that matter can be swallowed up in holes in space, has been verified. The detectors lifted into orbit around the Earth show that our Milky Way Galaxy is a violent place containing pulsars, X-ray stars, gamma ray sources, and a source of energy at the center of the

Galaxy itself, while we also see that further away across the Universe whole galaxies of stars are involved in giant explosions. Wherever we see cosmic violence today, theorists can explain events in terms of black holes, although when we begin to talk about violence on a galactic scale we need to invoke the presence of black holes containing hundreds of solar masses of material, or more (see Chapter 10).

But the revolution is not just in what we know of the Universe, but in the way we look at the Universe. Only twenty years ago, the basic philosophy underlying the study of the Universe was one of stability. The Galaxy was seen as a quiet place in which stars lived out their quiet lives in undramatic fashion, and the philosophy of continuity and steadiness extended to ideas of cosmology, the study of the Universe as a whole, where the idea of an unchanging "steady state" Universe, always looking much the same as it does now, became fashionable in the 1950s. This underlying view of the Universe, as much as anything else, is the reason why astronomers never thought to ask themselves the right questions that would have led to the prediction of pulsars, and had to be jolted out of their complacency by the accidental discovery of these energetic objects. Today, astronomers *expect* to find violence and change wherever they look in the Universe; the steady state theory is disproved, and the whole Universe is seen as having been created in a vast explosion, the Big Bang of creation. Galaxies suffer violent explosions and contortions; stars are tortured by gravity into explosive outbursts, as supernovae in the most extreme cases, leaving energetic pulsars and black holes behind.

And as the new philosophy of change filters back closer to home, the scales fall, metaphorically, from the eyes of scientists in many other disciplines. The Earth itself is seen as a changing, shifting pattern of continents, always moving and tugged around by the forces of continental drift. Our Sun, once thought of as a perfect, unchanging sphere of fire, comes under scrutiny in its turn, as astronomers now ask themselves why, if all around us we see change and violence, should our local star be immune from variations? The scale may be greatly reduced from the violence of stellar explosions, but then we are so much closer to the Sun that only relatively minor violence is needed to shake us up. Is the Sun, then, really a stable star on a scale that matters to human beings? Or is that balance between the inward tug of gravity and the outward pressure created by the heat of nuclear fusion in the interior susceptible to flickers that we might notice—perhaps fatally?

CHAPTER 7

# The Delicate Balance

THE STORY I'VE SPELLED OUT SO FAR REPRESENTS THE broad picture of stellar evolution. Astrophysicists are sure that they understand the broad outlines of the birth, life, and death of stars, and the general feeling is that since our Sun is "just an ordinary star" we can use the broad outlines of the standard picture to understand the workings of the Sun. But there is more to the story than this. It is much less openly admitted just how broad those outlines are. By and large, we can say that stars behave as outlined in the last three chapters. But all of this knowledge about stellar evolution comes from averaging thousands of stars; as I said before, there is no way we can sit and watch one star evolving for thousands of millions of years to test the theories in detail. And this is the big snag with our understanding of stars—averaging thousands of them to get "standard" patterns of behavior doesn't really show up small variations from one star to another, or between one moment in the life of a particular star and another moment a million (or a thousand million) years later. In particular, we cannot be sure, without direct measurements of some kind, that at this precise moment in its life our Sun is exactly in balance in

93

accordance with the theory of nuclear fusion, providing heat to hold the star up against the pull of gravity, as outlined above.

Without this inner heat, to be sure, the Sun would collapse. But it is so big and contains so much inner heat that it would not go out like a light. Turn off the nuclear fusion inside the Sun today, and it would take millions of years for the results to become apparent at the surface—and it is the heat from the surface that affects us directly. Or look at it another way—if the nuclear fusion *did* turn off less than 10 million years ago, we still wouldn't know it yet! All we can say is that *usually* the nuclear fusion reactions are "switched on," averaging over the 4,500 million years of the Sun's life on the main sequence so far.

There's another point that is often glossed over. We don't actually know for sure just how the nuclear reactions work. It's all very well to say hydrogen atoms stick together to make helium, and to mimic the reactions one by one by colliding beams of particles together in giant accelerators. But no one has ever been able to reproduce the conditions in the middle of the Sun in the laboratory (which is just as well, for the sake of the lab!) and *measure* the rates of the reactions, the temperature, and so on. Our knowledge of stellar interiors depends on extrapolating measurements made under less extreme conditions, using the best theories we've got; and the ultimate test of the theories is how well the resulting models mimic the behavior of the real Sun. In fact, the models produce a range of "suns" according to just how you tinker with reaction rates, the details of the convective region, and so on. There is still a range of uncertainty in our understanding of stellar interiors, down at the level of a few percent variation in this or that parameter. Nothing to worry about in terms of understanding the outline of stellar evolution, averaged over a thousand stars and a billion years; but a 10 percent reduction in the surface heat of the Sun would be ample to bring a complete ice age on Earth, so these nagging little uncertainties are a distinct cause for concern to you and me.

So these problems are more than just philosophical musings for the doubting Thomases who want to dot the i's and cross the t's of stellar theory. The possibilities are disturbing enough that even in the 1950s, when there was no hope of any direct measurements of what goes on inside the Sun, a few astrophysicists pondered on the implications of any short-term fluctuations in the workings of the solar interior, temporary disruptions of the delicate balance that might have repercussions for life on Earth.

The leading proponent of this kind of investigation at that time was Professor Ernst Öpik, then working at the Armagh Observatory in Northern Ireland. Starting out from the puzzle of the cause of ice ages, Öpik developed a model of the Sun in which during the normal hydrogen burning state heavy elements produced by the nucleosynthesis buildup around the core as a kind of "nuclear ash." Eventually, this "ash" becomes a barrier to radiation from the core, which can no longer get through effectively to the outer layers of the star. In such a state, the outer layers would shrink slightly, releasing heat from gravitational energy, while the "ash" would get hot and develop powerful convection currents which would then mix the heavy elements into the rest of the material of the star. At this point, the barrier around the core would be removed, heat could get out again, and the outer layers expand, returning the Sun to its normal state.

Today, more than twenty years later, astrophysicists have more observations to use in their calculations, and what they regard as better models of stellar interiors. They think, by and large, that Öpik's model of a Sun first choking in its own nuclear ash, then clearing its throat for another phase of normal nuclear burning, does not fit in with what we know about the behavior of stars, and has to be discarded, at least in its original

95

form. But at the same time that theorists have been expressing more confidence in their models, over the past decade or so a dramatic new observational development has made it possible to get the first direct measurement of processes going on in the heart of the Sun. And, to the embarrassment of the theorists, the new technique—neutrino astronomy—flatly contradicts their predictions and cuts the ground from under their confident feet. The result is that variations on the theme proposed by Öpik may, after all, be realistic approximations to what is going on inside the Sun today.

## THE NEUTRINO MYSTERY

We've already seen that electromagnetic radiation takes a very long time to get out from the middle of the Sun, and that in the process it is so scrambled up that no detailed information about the solar interior can be obtained from it. But there is a bizarre particle that is produced in nuclear fusion reactions—including the reactions which we believe keep the Sun hot—that hardly reacts at all with anything once it is produced. This is the neutrino, which can pass through the Sun itself, the solid Earth, or sheets of lead almost as if they were empty space; neutrinos produced by nuclear reactions in the solar interior should flood in their millions out through the Sun and across space, reaching the Earth only minutes after they begin their journey in the heart of the Sun. If we could only find a way of detecting them—taking the temperature of the neutrino "sea" —we would have for the first time a direct measurement of what is happening in the middle of the Sun, the ultimate test of all our theories.

This was the incentive which led Professor Ray Davis, of the Brookhaven National Laboratory, to begin work on a solar neutrino detector in the 1960s. The problem was to find a way to detect the elusive neutrinos without having a detector so sensitive that it would be swamped by the cosmic rays, energetic particles whizzing about in the region of the Earth. Davis's solution to the problem is to place his detector—a huge tank of fluid as big as a swimming pool—deep underground in a gold mine in South Dakota, so that the layers of rock above shield out everything except the neutrinos. Very, very occasionally one or two of the millions of neutrinos flooding through the detector every second should interact with atoms of chlorine in the fluid in the tank—and Davis has devised a way to count the number of chlorine atoms affected each month.

The details of the whole business are vitally important, and have been puzzled over by the top brains of astrophysics and nuclear physics for more than ten years—for, it turns out, the Davis detector does not find anywhere near as many solar neutrinos as theory predicts. Has the Sun's nuclear furnace switched off, after all? The implications are so enormous that, clearly, it is worth looking at the physics in detail, starting from what we already know about the nature of the Sun.*

*Figures 13–15* courtesy of *Engineering & Science,* California Institute of Technology, and Professor W. A. Fowler.

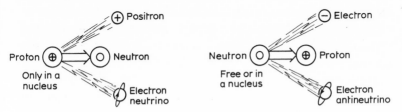

*Figure 13* Neutrinos are produced when protons are changed to neutrons; antineutrinos are produced when the process is reversed.

*The following description of the neutrino astronomy technique is a shorter version of an article of mine which appeared in *New Scientist* 80 (1978): 780.

97

The Sun is mainly composed of hydrogen—or rather, hydrogen nuclei (protons) swimming in a sea of electrons, since the temperatures involved are far too great for nonionized atoms to exist. Because of the rapid motion of the nuclei at these high temperatures, two protons can collide and fuse into a deuteron (the nucleus of deuterium) with one of the protons being transformed into a neutron, liberating a positron and a neutrino in the process. This is a proton-proton (p-p) neutrino; but the story doesn't stop there. Deuterons can also fuse with other nuclei, and may combine with a proton to form a nucleus of helium-3 (two protons plus one neutron), with no extra neutrino produced at this stage. Helium-3s do not combine with single protons, as laboratory experiments show, but they can join together in pairs with spectacular results.

When two helium-3 nuclei collide, two energetic protons are released, and two protons and two neutrons stick together as a helium-4 nucleus. The net result is that four protons have been converted into one helium-4 nucleus, energy has been liberated (keeping the Sun hot) because the mass of helium-4 is slightly less than that of four protons, and two neutrinos have shot off through the Sun and out into space, while two positrons are left in the Sun. Helium-3 and helium-4 can also combine to form beryllium-7, with no neutrino produced, and the beryllium-7 can capture electrons, changing one proton into a neutron to become lithium-7 and emitting another neutrino; this is much rarer than the helium-3/helium-3 interaction, but does happen occasionally. Eventually, the lithium-7 produced captures a proton and splits into two helium-4 nuclei.

Even more occasionally, beryllium-7 captures a proton, gives off a neutrino, and becomes boron-8; this decays in a little over a second to two more helium-4s, giving off a positron and, yes, another neutrino. This may seem a long and devious tale (although it is only the very beginning of the road of nucleosynthesis by which everything has been made); but it is directly rele-

98

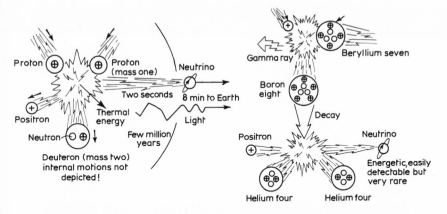

*Figure 14* The two processes which astrophysicists believe ought to be making neutrinos inside the Sun: left, the proton-proton process; right, beryllium and boron interactions.

vant to the present solar neutrino puzzle. For, although the neutrinos produced earlier in the chain (the p-p neutrinos) are by far the most common, they are not very energetic and therefore they are difficult to detect. The boron-8 decay neutrinos, by contrast, are much rarer birds—but are very energetic and much easier to detect, in principle. So these are the ones that Davis set out to find.

First, though, he had to have some idea of how many to look for, and this is where the experimental nuclear and particle physicists come in. By accelerating protons, deuterons, helium-3 and helium-4 nuclei in beams and shooting them at various targets, the experimenters determine the rate at which the different reactions occur, and how the reaction rates change with temperature. They then extrapolate these measurements to the conditions appropriate for the inside of the Sun, and deduce how many neutrinos of each kind ought to be produced. The next problem is to catch them. This Davis does with the aid of a reaction involving chlorine-37, and since chlorine is not a very nice gas to handle he keeps it locked up in perchloroethylene ($C_2Cl_4$), cleaning fluid. Conveniently, on average one chlorine atom in four is the isotope 37, and there are four chlorine atoms per molecule of perchloroethylene, so each molecule should have one chlorine-37 atom and the total number in the tank can be determined.

When a sufficiently energetic neutrino collides with a chlorine-37 nucleus, it knocks out an electron, turns one neutron into a proton, and the atom becomes argon-37. This is the reverse of the way the neutrino was made, when a proton became a neutron, releasing a positron and a neutrino; it is important that the neutrino involved should be the kind associated with electrons and positrons, and although other kinds of neutrinos exist they will not trigger this transformation, however much energy they have.

The argon-37 is a gas which escapes into the fluid around it and then decays, with a half-life of thirty-five days, by capturing an electron, ejecting a neutrino, and changing back to chlorine-37. This would be rather pointless—but there is one more subtlety. The chlorine-37 now produced is in an energetic state, and in restoring normality it ejects an electron with a characteristic energy. This electron, with a characteristic energy of 2.8 KeV, is what Davis detects with the aid of special counters; *each* such electron can be counted, and each corresponds to exactly one energetic electron neutrino interacting with the material in the tank. The tremendous technical difficulties of flushing the argon out of the tank once a month and actually counting the electrons, and of making tests with radioactive argon and chlorine from other sources to show the system works, deserve more recognition than Davis has yet achieved, and seems to be the most solidly founded step in the whole chain.

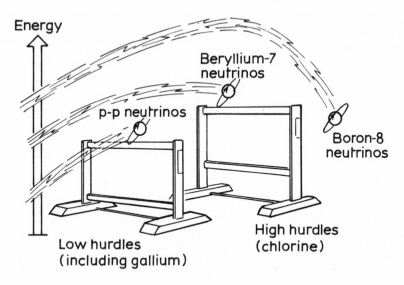

*Figure 15* Low and high hurdles for neutrinos. The p-p neutrinos are less energetic than beryllium-7 and boron-8 neutrinos, and a lot of energy is needed to trigger the chlorine-based detector used by Ray Davis and his team. Though much more numerous, p-p neutrinos simply can't get over the energy barrier to be detected. If a new detector is built using gallium in place of chlorine, astrophysicists may be able to measure directly the flux of p-p neutrinos and solve the solar neutrino problem. But the detector would need sixty metric tons of gallium, which costs $520,000 a ton today (1979), with world production only seven tons a year!

Now the two ends of the chain have to be matched up.

How many solar neutrinos should the detector detect, in theory? The standard model of the Sun leads to a prediction of a juicy 60,000 million p-p neutrinos per square cm per second, flooding through space and, unperturbed, through the solid Earth as well. Alas, *none* of these can be detected by Davis since they do not have enough energy to trigger the chlorine/argon switch. The beryllium-7 neutrinos have just enough energy, but

101

are much scarcer; and the boron-8 neutrinos can easily do the job, but are scarcer still. But the exact numbers of each of these —unlike the p-p neutrinos—is very dependent on the temperature assumed by the model for the center of the Sun.

With a central temperature of 15 million K, the beryllium-7 neutrino flux is 4,000 million per sq. cm per second and the boron-8 neutrino flux is 3 million per sq. cm per second. The former should produce just five captures in Davis's 100,000 gallon tank each month, and the latter, lesser but more energetic, twenty. The expected total, after all that effort, is just twenty-five captures per month—a dramatic indication of just how reluctant neutrinos are to interact with anything, and how many skip through the detector each month undetected. And in the 1970s in over thirty test runs, the actual totals observed averaged out to just nine captures per month. In round terms, just one third of the expected number of solar neutrinos are being detected—so what has gone wrong?

Maybe we don't understand the nuclear physics, after all. Or, just maybe, we don't understand the Sun as well as we thought. All you need to get a "prediction" which agrees with Davis's measurements is to lower the central temperature in the model in the computers by about 10 percent, and there are many ways of tinkering with the standard model to achieve this. One way is to set the mixture of heavy elements in the Sun slightly richer than in the standard model, though that raises the puzzle of how they got there in the first place. Some astrophysicists, rather apologetically, have revived Öpik's idea of a more or less regular "mixing" of the Sun's interior, usually making a point of saying in their scientific papers that they don't really believe the models they are putting forward, even though they do seem to explain the solar neutrino puzzle. For, of course, anything which has disturbed the normal life of the Sun within about the past 10 million years could have produced a temporary switch-off in the nuclear reactions and slight cooling.

## LINKS WITH ICE AGES—AND THE GALAXIES

Now the puzzles come thick and fast. Is it any coincidence that for the past couple of million years the natural state of the Earth has been a full ice age, and warm spells like the one we are now living through (interglacials) have been few and far between? If the Sun is temporarily off color, what happens when it switches back on and things return to normal—the hothouse Earth the dinosaurs reveled in tens of millions of years ago? The death of the Sun may be a long way away; but now, it seems, there may be factors in the life of the Sun for us to worry about too.

One idea revived recently in the light of these puzzles suggests that from time to time in its orbit around the Galaxy the Solar System encounters patches of interstellar material, dust and gas clouds in space. The shape of our Galaxy, remember, is a flattened disk with a spiral pattern marked out by two twisting "arms" of bright stars edged by dark clouds of dusty material. The bright stars are born out of the dark clouds, squeezed in successive passages through the spiral shock wave (see Chapter 4). But, of course, the Sun and Solar System continue in the same orbit even after they have formed, and on every circuit of the Galaxy they encounter the dust clouds and shock wave of each spiral arm. Several people hit on the same idea more or less simultaneously in the 1970s when they put this viewpoint together with the geological evidence that great ice epochs on Earth occur roughly every couple of hundred million years. Are the intervals between the passages through spiral shocks right to account for ice epochs in terms of an effect on the Sun?*

---

*Actually the idea is an old one which now looks more plausible than ever before; the full historical credit is given by W. H. McCrea in his article, "Ice Ages and the Galaxy" (see bibliography).

Various ideas have been tossed in the ring—perhaps the dust simply acts to block out solar heat, cooling the Earth. More bizarrely, some theorists suggest that the dust falling into the Sun might make it hotter, and that a hot Sun could produce an ice age on Earth by a devious interaction. The argument runs that as the Sun warms more water evaporates from the oceans, forming a dense blanket of clouds around the Earth; this then reflects away most of the Sun's heat, so that the Earth below cools and all the water in the atmosphere falls as snow. Then, as the air clears, the snow takes over the job of reflecting away solar heat and, bingo, you've got an ice age.

Once again, nobody seems really to take the models seriously (especially not the hot Sun/cold Earth model!). But what are the effects on the Sun's interior going to be when it passes through a dust cloud and the natural radiation balance of the surface layers is disturbed by obscuring matter? In all honesty, we don't know. The problem is that anything disturbing the radiation balance of the outer layers—such as gravitational energy of infalling matter being turned into heat—is going to disturb the convective layers of the star. And when I put this point to one eminent astrophysicist involved in the study of stellar structure, he replied, "Frankly, nobody knows *anything* about convection." (This slight exaggeration means that all we know about convection is the broad principle that it does happen; we don't know anything about the details of exactly under what conditions it switches on or off.) Another astrophysical modeler offered a little more constructive help, showing me calculations which seem to give a good idea of what happens when the material is falling onto the star (the radius increases a little bit, the central temperature drops by a tiny amount, and the star cools off marginally), but he then pointed out that the tricky question is what happens when the new source of energy is turned *off.* You can make a computer model which works with the infalling matter and reaches a new stable state; but take the dust away and the model runs wild (in technical terms, it "fails to converge" on a new stable state).

This tells us two things. First, obviously, the models aren't good enough. But, secondly, the fact that they "fail to converge" under these conditions shows that something drastic happens to the star. So, with no backing from computer models because they can't do the job, I offer here a possible "model" based on physics and common sense.*

By and large, the infalling material inhibits convection, by reducing the temperature difference between the surface and the interior. It's a bit like a pan of water boiling on a stove; with a lid on the pan the bubbling convective fluid reaches an "equilibrium" state and all is well, but when the lid is lifted off there is a brief flurry of extra activity as some of the pent-up heat from the furnace below is released. Then, of course, the bubbling pan settles into a new "equilibrium." So if infalling material inhibits convection, removing the infalling material must suddenly let convection run wild for a brief time—brief, that is, by comparison with the Sun's lifetime. The sudden extra mixing can easily change conditions in the interior sufficiently to produce a 10 percent drop in temperature, as Öpik showed back in the fifties, and it could easily take 10 million years or more for the Sun to recover from such a convulsion and return to normal nuclear burning.

Depending on what else happens to the Earth in the meantime, our descendants may be able to find out firsthand just how accurate these simple analogies are, since some recent observations suggest that the Solar System is even now heading toward an interstellar dust cloud, and will plough into it in about 50,000 years. But, much more important for the present understanding of the absence of solar neutrinos here and now, we know that the Solar System has indeed recently crossed a dust

---

*First offered in my book *White Holes* as an afterthought!

105

lane, edging the so-called Orion arm of our Galaxy, and may have emerged from the Orion Nebula itself only a few tens of thousands, let alone a million, years ago. *If* this rather vague and speculative association between dust, the Sun, and solar neutrinos means anything at all, then here and now is exactly where and when the Sun ought to be temporarily off color, recovering from the throat clearing that occurred when it emerged from the dust.

## A SHAKING SUN, AND OTHER ODDITIES

But this is still not the end of the story of the accumulating evidence of solar peculiarities. One little tidbit comes from spectroscopic studies of the Sun's atmosphere made by a team from Hawaii University using rocket-borne detectors carried high above the Earth's atmosphere. They found, in 1977, that whereas meteorites contain nineteen times more tin than they do of the element indium, the Sun's atmosphere has more indium than tin. This is puzzling because both meteorites and the Sun's atmosphere "ought" to show element abundances roughly in line with the composition of the original cloud from which the Solar System formed—yet there is ten times more indium in the outer layers of the Sun than our models account for. Another little chip at the acceptability of those models.

Meanwhile, following the neutrino saga, astronomers have found another way to get a handle on what happens inside the Sun with the discovery that the whole Sun wobbles like a jelly —the solar equivalent of earthquake vibrations. Detecting these tiny oscillations is another triumph of observational technique, and once again it leaves the theorists with egg on their faces. The discoveries show several intermingled oscillation "modes," one with a peak velocity of only 2 meters per second, having a period of 2 hours and 40 minutes. Just as the speed with which earthquake shocks travel through the solid Earth depends on the structure of the Earth, giving geophysicists a tool to probe the interior of our planet, so the exact period of these solar oscillations depends very precisely on the detailed structure of the Sun, including its density and temperature. Of course, the modelers have come up with a whole heap of models "explaining" the oscillations; but one of the leading contenders for explaining the 2 hour 40 minute vibration is a standard solar model with only one change—the temperature of the interior has to be reduced by about 10 percent.

Once again, the theorists say that the case is not proven one way or another. There are other ways to juggle the figures to get this kind of oscillation, although they involve rather more juggling than this simple change. Yet all of the theorists are uncomfortably aware that this one change, reducing the core temperature by 10 percent, *exactly* explains both the neutrino puzzle and the observed sunquake phenomenon. Well, you pay your money and you take your choice. But my choice is definitely for *one* explanation which explains *both* puzzles, since otherwise we need two different theories, each of them more complicated than the cool interior model, to account for the two observations separately.

Every time we get new information about the *details* of solar behavior, rather than the broad outline, it points the same way —the standard model is just plain wrong, and the Sun is not in a typical state of main-sequence nuclear burning today. The delicate balance in the interior has been disturbed in some way (maybe by passage through a dust cloud, maybe not). How far can such a disturbance affect the surface of the Sun, and thereby conditions on planet Earth? The conventional wisdom is that, buried beneath the layers of the convective and radiative zones, and with the buffer of gravitational energy available from slight contraction or expansion of the Sun's radius, minor changes in the core need not affect the surface much at all. But by now we should be pretty wary of the conventional wisdom—and, dead on cue, in the very week that I was preparing the material for this chapter, another blow to conventional wisdom regarding stellar interiors was struck by a brief scientific paper in the journal *Nature*.

This paper* reported a comparison between the solar neutrino measurements, which provide a measure of conditions in the deep interior of the Sun, and observations of changes in the numbers of spots on the surface of the Sun. These sunspots are generally regarded as features associated with magnetic fields and electric currents in the surface layers of the Sun, with little or no connection with events in the deep interior. They come and go over various cycles of activity, the best known being the roughly 11-year-long "sunspot cycle" characteristic of all variations in the activity of the Sun's surface and surrounding magnetic fields. But on top of this main pattern of variation there is a lesser ripple, only recently identified, just over two years long; it is a slight tendency for the numbers of sunspots to wax and wane with a period (or quasi-period) of 25.3 months. This newly discovered effect has only been measured and analyzed so far for the years 1970–76; but by a happy coincidence this is just when Davis's solar neutrino experiment was really getting into full swing. And when we make the same kind of analysis of the very small variations in the neutrino counts over the years 1970–76, above and below the average level of nine counts a month, what do we find but a quasi-period of just over two years, exactly following the sunspots' 25.3 month variation!

This, surprise, surprise, is not something that can be explained by the standard model of the Sun, suggesting that some unknown process (perhaps related to the convection which "nobody knows *anything* about"?) inside the Sun links changes affecting the deep interior with the very surface itself. This is

*By Kunitomo Sakurai; see bibliography.

obviously going to be an interesting area of research in the next few years, but I am more than happy to leave the puzzle as it stands now, with a large question mark. For I had already intended to develop my theme of solar variations, a disruption of the delicate balance, by looking at the intriguing puzzle of sunspots and their variations over past centuries and millennia. I had thought it might be difficult to justify such an abrupt jump from the deep solar interior, the heart of the neutrino puzzle, to the surface layers and the puzzle of sunspots. But now, it seems, nature has done the job for me—there *is* a connection between the nuclear burning processes in the core and the sunspots on the surface, so what could be more natural than to turn from changes in the interior to changes on the surface, the more obviously visible flickering manifestations of the inconstancy of our Sun?

# CHAPTER 8

# The Inconstant Sun

THE TIMESCALES THAT MATTER TO MAN ARE THOSE OF decades, centuries, and, at the most, millennia. What happens over the next ten years is of vital importance to all of us; with a large global population and complex technological society, some plans now being made by governments and international agencies (construction of dams, roads, and other big projects) will have repercussions for several decades at least, affecting the lives of our children; and both historians and those concerned with the long-term future of mankind must be uncomfortably aware that the environment here on Earth, especially the climate, changes considerably on a timescale of a few centuries. Changes in the output of heat from the Sun, or in the nature of the radiation it emits, will have profound effects if they occur on any of these timescales. And, indeed, some climatologists now suspect that the climatic changes of the past thousand years or so can be explained by solar variations.

But now we are talking about the merest ripples on the surface of the Sun, a situation quite removed, it was thought before 1979, from anything going on in the deep interior. Then, in addition to Dr. Sakurai's astonishing discovery of a link between sunspots and solar neutrinos, Professor Robert Dicke, of Princeton University, came up with independent evidence that the behavior of the spots on the surface of the Sun is modulated by processes in the deep interior. As he puts it, there seems to be a chronometer hidden deep in the Sun. And all this becomes of crucial importance once we realize that those spots on the solar surface, along with other features of activity over the visible face of the Sun and out into space, vary on just the timescales that matter to man, over decades, centuries, and millennia.

What is a sunspot? In essence, it is simply a relatively cool region of the Sun's surface, appearing dark by comparison with the even hotter material around it—a mere 8,000 degrees or so instead of 9,000 degrees!* Individual spots may be anything from about 1,500 km across to 150,000 km, and they are associated with strong magnetic fields which may cool the region by inhibiting the all-important convection. Spots are simply the most visible manifestation of a whole range of solar activity—when there are more spots, the Sun is more active, producing a variety of phenomena related to solar magnetism, great flares which blast material up and out into space, disturbing the solar "wind" of particles which continually streams out past the orbits of the Sun's family of planets. The whole pattern of activity waxes and wanes over a cycle roughly 11 years long, with the solar magnetic field being reversed after each cycle so that it gets back to its starting state after 22 years. This magnetic cycle is generally regarded as the basic pattern of variation

*Where you define the "surface" of the Sun depends upon your viewpoint. Looking in from outside, we see the active spot regions at a depth corresponding to about 9,000 K, but a standard definition of the surface might be a little above that, where temperature has dropped to around 6,000 K. All definitions —like the definition of the "edge" of the Earth's atmosphere—must be more vague than a definition of the solid surface of a planet.

around which all the other bits and pieces, including changes in the number of sunspots, should be placed. But although modern techniques, including observations from satellites above the Earth's atmosphere, show that the whole of the surface activity of the Sun is undergoing this regular upheaval, with the changing number of spots simply a very minor symptom, like the rash on the skin of a measles victim, it is still common to call the pattern by its old name, the "sunspot cycle."

The basic 11-year sunspot cycle was only recognized about a hundred years ago, and ever since then it has been generally regarded as a very rough and ready "cycle." The spacing between years of peak activity has sometimes been as little as 7.3 years, and sometimes as great as 17.1 years, making a pretty vague "chronometer" and causing astronomers to puzzle over whether there is any "real" cycle or simply a succession of independent disturbances each of which take about 11 years to die down.

Professor Dicke's study completely transforms this view. He used all the power of modern statistical methods to ask what is in fact a very simple question: Do the changes in the length of individual cycles occur at random, or do they add up over the centuries to give a clearer pattern? And what he found is that, taking many cycles together and "averaging" them in the proper statistical sense, it seems that the sunspot numbers are being kept very precisely in step by some great solar metronome. One example spells out how this works. Starting in 1761, the historical records* show that there were three very short cycles averaging 8.9 years, at the end of which the year of peak solar activity arrived 5.6 years "early" compared with the 33-year interval that "ought" to have occurred. But the next cycle

---

*Although the sunspot cycle was only recognized in the late nineteenth century, recorded observations of sunspots go back much further.

113

was 17.1 years long, putting the sunspots exactly back in step with the 11-year rhythm, after an interval of 27 years and more. "It is as though," says Professor Dicke, "the Sun 'remembered' the correct phase for 27 years and then suddenly reset the sunspot cycle."

The story is the same right down the centuries, once it is seen with the powerful insight provided by statistical analysis of the records. *Individual* sunspot cycles may be long or short, but the average is kept very accurately in step.

There is only one reasonable explanation of such a phenomenon. It must be that something deep in the Sun is pulsing with a precise 11-year (or 22-year) period, and that *all* the surface effects of the solar cycle of activity are simply an imperfect reflection of this deep-seated pulsation. The "errors" in the variation of the surface features must be due to imperfections in the way the "message" gets transmitted to the surface—it may take several years for the beats of this solar heartbeat to spread their influence outward, and sometimes it may take more years than at other times. It is far too soon yet for the theorists to have come to grips with this new evidence and to produce a better understanding of how the Sun works. That will take years, involving contributions from studies of oscillations of the solar surface, neutrino astronomy, and observations from orbit across the electromagnetic spectrum. But it is already further evidence of the variability of our own star, the Sun. And it provides appropriate background to look at the relationship between the solar cycle of activity and events here on Earth.

## HOW VARIABLE IS THE SUN?

From the first Victorian astronomers who noticed the sunspot cycle down to the present day, people have tried to find links between the 11- and 22-year solar rhythms and the weather on Earth. Many claims have been made, involving claims of such patterns in rainfall figures for England, changing water levels in African lakes, and the number of times lightning strikes each year. Some extroverts have gone so far as to link the 11-year cycle with changes in stock market prices; but it has proved impossible to explain all of these claimed correlations in terms of a genuine solar-terrestrial link. The balance of evidence is that the link is real,* but just as the deep-seated solar rhythm gets distorted by the time it reaches the surface, so the roughly cyclic changes in the surface activity of the Sun get distorted and diffused in their influence on the Earth. One hint of why this might be so comes from a combination of spacecraft observations and ground-based measurements.

For a hundred years, astronomers have been teased by the puzzle of whether the actual amount of heat being radiated by the Sun varies over the sunspot cycle—in other words, is a more active Sun hotter? Living at the bottom of the Earth's atmosphere, we simply can't make measurements of the "real" output of the Sun (the so-called "solar constant") because an unknown amount of solar radiation is absorbed by the atmosphere before reaching our instruments. The ground-based measurements do suggest that the solar "constant" changes, by as much as 1 percent on a very short timescale linked with the lives of individual spots. But now instruments mounted on spacecraft such as Mariner 6 and Mariner 7 show no evidence of changes even one tenth as big as this due to the sunspots themselves, on the appropriate timescale of a few days or weeks. So something must be happening to the solar radiation as it passes through the Earth's atmosphere, and in particular it seems that since more ultraviolet radiation is produced when the Sun is active, the ozone concentration high in the stratosphere changes, altering the effectiveness of this layer as a barrier to the Sun's radiation.

*See *What's Wrong with Our Weather?*

115

All this tells us nothing about whether the solar "constant" changes by 1 percent or so from decade to decade, or between one century and the next—the measurements only looked at short-term effects of specific sunspots and associated activity. But it does hint very strongly that the atmosphere is a variable filter, sometimes letting more heat through to the ground than at other times. This alone is bound to produce an averaging effect, smearing out any solar cycles by the time their influence penetrates into weather systems. And this smearing effect is why it makes no sense to talk about changing patterns of weather and climate linked with changes in solar activity on any timescale less than a decade—roughly the average over a sunspot cycle.

If there is one burst of solar activity in one month of one year, the climatic patterns of the Earth won't be changed (although parts of our planet may feel the jolt; see Chapter 9). But if the Sun is either very active or very quiet for decades or centuries at a time, then we might expect to see the effects as broad, long-term changes in the climate of the Earth. One nagging piece of evidence provided the impetus for this possibility to be taken seriously. According to all the astronomical records we have, the seventeenth century was a period of greatly reduced solar activity, with very few spots at all seen for more than fifty years. This period exactly coincided with what climatologists know as the worst period of the "Little Ice Age," when winters were so harsh that many European rivers such as the Thames froze, glaciers advanced, and the repeated failure of crops brought recurring famines. Throughout the twentieth century the question has been asked: Was there really a dearth of sunspots during the Little Ice Age, or is it simply that nobody bothered to keep accurate records in the seventeenth century? In the mid-1970s the question was resolved by Dr. John Eddy, of the National Center for Atmospheric Research in Boulder,

Colorado, who turned detective-historian to answer the riddle plaguing climatologists.

## THE MAUNDER MINIMUM

Sunspots were known to the ancient Greeks, but this knowledge was lost in the West and the spottiness of the Sun only rediscovered by Galileo in the early seventeenth century. (The Chinese knew about sunspots throughout this period when no observations were being made in the West—but more of that later.) The realization that the Sun shows blemishes on its skin was among the "heresies" for which Galileo was persecuted; the central dogma of the Western Church in the centuries prior had been that the Sun was a perfect sphere, created in its perfection by God, so that suggesting imperfection in the Sun was seen in some quarters as implying that God could be fallible and was guilty of shoddy workmanship. Nevertheless, the facts couldn't be covered up once the rediscovery of sunspots was made, and indeed the religious arguments attending Galileo's observations should have ensured a keen interest in the Sun in the ensuing decades. Could the Church have suppressed written evidence of sunspot observations? Or could people have lost interest so quickly that no one bothered to publish news of such observations in the second half of the seventeenth century? Eddy's analysis of the historical record shows not; astronomers eagerly sought sunspots at this time and reported them at length when they appeared. They just didn't appear very often.

Today, even at the minimum of the sunspot cycle we might see half a dozen spots in the course of a year, while during the year of maximum solar activity in the cycle we might see 100 or more (such a peak of activity is due in 1980 or 1981). Since the seventeenth century, there has been only one year (1810) in which no sunspots were recorded; but after the realization that

117

there is an underlying period to the sunspot cycle, Victorian astronomers of the 1890s tried to find how far back this period could be traced, and searched the records right back to Galileo's time. Pioneers Gustav Spörer and E. W. Maunder found that almost no sunspots were seen from about 1645 to 1715, a period dubbed the "Maunder Minimum" by modern astronomers. Between them, Spörer and Maunder published five detailed scientific papers on the subject in the years up to 1922, drawing on evidence such as the report in the *Philosophical Transactions of the Royal Society* in 1671, where the editor commented:

> At Paris the excellent Signior Cassini hath lately detected again Spots in the Sun, of which none have been seen these many years that we know of.

Cassini's own report of the phenomenon included the comment:

> It is now about twenty years since astronomers have seen any considerable spots on the Sun, though before that

*Figure 16* Changes in solar activity since the seventeenth century show up clearly in this plot of the changing sunspot number from year to year (the number plotted is the monthly mean sunspot number for the appropriate year).

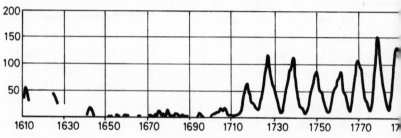

118

time, since the invention of the telescopes, they have from time to time observed them.

And the then Astronomer Royal, John Flamsteed, was so excited by the sighting of a spot in 1684 that he commented:

This is the only one I have seen . . . since December 1676.

We can surely take it for granted that the Astronomer Royal was *looking* for sunspots in the intervening eight years! But, in this modern age, astronomers have been curiously reluctant to accept Spörer's and Maunder's accounts. There is still too much of an arrogant tendency to dismiss nineteenth-century scientists as bunglers, even though our present scientific knowledge builds from their work. So it was necessary for Eddy to silence these doubting Thomases by going back to the original sources, repeating and improving on the studies of his two pioneering predecessors, and publishing his results in the journal *Science* in 1976, a respectable journal and a modern date to bring the evidence to the attention of late-twentieth-century science.

His research is impeccable and his results unquestionable. The Maunder Minimum is real; sunspot activity really did

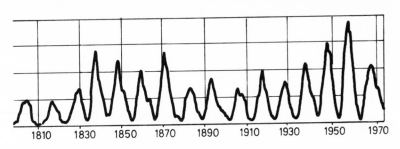

119

almost switch off between 1645 and 1715. Not only the sunspot records show this; when the Sun is more active the gusts of the solar wind spill charged particles into the Earth's magnetic field where, focused at the poles, they produce those great free light shows, the aurorae. More and brighter aurorae mean more solar activity; and, significantly, very few bright auroral displays were seen in the second half of the seventeenth century. The solar wind also has another effect on the environment of our planet. When the Sun is active and the solar wind is strong, it shields us from cosmic rays from interstellar space. But when the Sun is quiet and the solar wind is weak, these cosmic rays penetrate the atmosphere in great quantities. There, they can react with nitrogen atoms to make atoms of the isotope carbon-14, the radiocarbon so useful in dating samples of old trees and other vegetation. Tree rings can be dated accurately simply by counting the layers of wood in a sample, and there are many trees around that are three or four centuries old. Each ring contains carbon (wood) which includes a proportion of carbon-14 depending on the age of the ring (which is known) and the strength of the cosmic rays in the year that the wood was growing. So tree ring studies give an accurate measure of the amount of cosmic radiation penetrating the solar wind each year—in other words, they tell us how active the Sun was each year. This very modern technique reveals the Maunder Minimum, with great clarity, as a period of *increased* atmospheric carbon-14 production, caused by a *decrease* in solar activity and in the strength of the solar wind.

So the Maunder Minimum is real and we have proof that the Sun can vary in a way which directly affects the Earth (certainly through carbon-14, perhaps by affecting the climate) for periods of tens or hundreds of years.*

*Quotations in this section are as reported by Eddy, *Science,* 192 (1976): 1200.

With proof that the Sun is variable in an irregular way, and with the new tool of radiocarbon analysis of old wood samples as an aid, the past few years have seen astronomers delving back into the historical records, from both East and West, to find out just how much solar variation there was in the centuries before Galileo. The Western records provide circumstantial evidence from reports of aurorae; but the Eastern records include mention of sunspots themselves, as well as aurorae and other effects. The pattern that emerges is dramatic indeed for anyone who still suffers from the delusion that our Sun is either perfect, constant, or regular.

## BACK TO THE BRONZE AGE

With the new carbon-14 tool proved as a good guide to solar activity, John Eddy has been able to use the technique to push back the story of solar variability to the Bronze Age. During the past 5,000 years, he finds, there have been times when the Sun was much more active than it is today, and also times when it was much less active. He describes the 11-year cycle as "but a ripple on an ocean of great and sweeping tides," tides which have brought at least twelve major "excursions" of solar activity away from the pattern we think of as normal since the Bronze Age. The carbon-14 record extends the study of solar variations back to 3000 B.C., about halfway between the present and the end of the most recent ice age here on Earth, and far beyond the scope of the written record.

Eddy uses the Maunder Minimum as his yardstick in assessing these events, defining it as an "excursion" of magnitude $-1$ compared with the present day; the state of the Sun now, on this broad picture, is seen as moving toward a new maximum of activity, on a scale of a hundred years or more, the like of which has not been seen for many centuries, since the "Medieval Maximum" of the twelfth to fourteenth centuries. Using Eddy's names for the various excursions, and his scale in which the Maunder Minimum defines a unit of $-1$, the twelve major events since 3000 B.C. are given in Table 1.

There is no evidence that these long-term changes in solar activity follow any cyclic pattern, and indeed clusters of positive excursions (more solar activity) and negative excursions (less solar activity) seem to be the rule, rather than a switch from positive to negative or vice versa. The two quiet periods of the millennium before Christ must have been dramatic indeed—each twice as extreme as the Maunder Minimum and together lasting for 360 years, more than a third of that millennium. Like the Maunder Minimum period of the seventeenth century, these were cold centuries on Earth, further striking evidence that when the Sun is less active the Earth cools down. But on this sort of timescale, rather than the monthly or yearly flickers of individual bursts of sunspot activity, it seems that the temperature of the Sun itself may change. It's not that the temperature changes follow the sunspot cycle of activity;

TABLE 1

Extreme periods of solar activity since 3000 B.C.

| Event | Duration | Magnitude |
|---|---|---|
| Sumerian Maximum | 2720–2610 B.C. | $+1.3$ |
| Pyramid Maximum | 2370–2060 B.C. | $+1.1$ |
| Stonehenge Maximum | 1870–1760 B.C. | $+1.3$ |
| Egyptian Minimum | 1420–1260 B.C. | $-1.4$ |
| Homeric Minimum | 820–640 B.C. | $-2.0$ |
| Grecian Minimum | 440–360 B.C. | $-2.1$ |
| Roman Maximum | 20 B.C.–A.D. 80 | $+0.7$ |
| Medieval Minimum | A.D. 640–710 | $-0.7$ |
| Medieval Maximum | A.D. 1120–1280 | $+0.8$ |
| Spörer Minimum | A.D. 1400–1510 | $-1.1$ |
| Maunder Minimum | A.D. 1640–1710 | $-1.0$ |
| Modern Maximum | Began nineteenth century–still growing? | |

rather, the strength or weakness of the sunspot activity is a *result* of changes in the Sun associated with temperature changes. In 1976, Eddy described this viewpoint as a "hunch"; the discovery three years later of links between sunspot activity and deeper, buried processes inside the Sun must make the idea far more respectable than that. No wonder two generations of astronomers failed to detect evidence that sunspot changes cause changes in the temperature of the Sun—their investigations are described only too aptly as looking for a means by which the sunspot tail could wag the solar dog!

The peak range of variations in solar output over the past 5,000 years need only be about 1 percent to produce changes in temperature on Earth averaging 1 or 2 degrees, ample to account for all the climatic changes since the Bronze Age, but these leisurely changes over several decades would be very hard to measure directly. Even so, some measurements hint that solar output increased by about 0.25 percent in the first half of the twentieth century, a time of increasing solar activity when each sunpot cycle in turn was more active than the one before. But while the carbon-14 story makes a complete and convincing tale, such concepts are still so startling to astronomers brought up to believe in the dogma of solar constancy that it is well worth emphasizing that this is not the only record we have of changes in solar activity over the millennia.

## ORIENTAL CONFIRMATION

Whereas Western philosophers from the time of Aristotle to the seventeenth century were handicapped by the dogma of perfection and constancy in the Universe, Eastern scholars have always been much more interested in the changing aspects of the heavens, phenomena such as comets, new stars (novae and supernovae) and, of course, the changing face of the Sun. They also had the right conditions for viewing sunspots without the aid of telescopes. As every astronomy primer stresses, it is dangerous to look at the Sun with the naked eye, let alone directly through a telescope (which is why sunspot observations today are made by projecting the telescopic image on a white screen), and you certainly won't see sunspots, even if they are there, by staring at the blazing Sun at noon. Some natural disturbance of the atmosphere is needed to dim the light from the Sun at the same time that spots large enough to be seen with the naked eye are around. Clouds are no good—they block out the Sun altogether. So what you need is haze in the atmosphere, or a dust storm, especially effective at sunrise or sunset when the Sun is low on the horizon and its radiation passes through a longer column of air than at noon.

These conditions occur much more frequently in the heart of a continent or, with prevailing winds from west to east, on the eastern edge of a continent where the winds are coming off the land rather than the sea. So England and western Europe are distinctly bad places for catching a naked eye glimpse of a sunspot, while China and Korea are ideal. Hardly surprising, then, that modern solar astronomers such as Dr. David Clark, of the Royal Greenwich Observatory, have turned to Oriental records for their studies of the changing Sun—and hardly surprising, too, that when those records do mention sunspots they very often comment at the same time that the brightness of the Sun was reduced. This does *not* imply that the Sun itself was dim, but that dust or haze in the atmosphere was obscuring its light and making it possible to pick out the spots against the glare of the solar disk.

These records go back in China in reliable form to the Han Dynasty (200 B.C. to A.D. 200), by which time a well-organized astronomical/astrological bureaucracy was established with the task of keeping the Emperor informed of portents from the heavens, and responsibility for maintaining the calendar. The historical records of each succeeding Chinese dynasty include a wealth of astronomical observations, and from the tenth century similar records are available from Korea. Generally, the astronomer/astrologers of the time regarded sunspots as a sign that something was up, but they were far from consistent in deciding just what such an omen portended; it is also quite possible that they didn't bother to mention all of the available "signs from heaven" in years when the Emperor seemed in no need of heavenly guidance. David Clark cites the example of the Chin Dynasty (A.D. 265 to 420) when very few "warnings from heaven" of any kind were published in the records just after the new Emperor came to power, when all seemed well with the world. Later in the dynasty, as dissatisfaction with the regime grew, a great many more heavenly events are put into the record as the then Emperor's celestial advisors struggled to find some comforting advice for him. So the astronomer/historian needs also to be aware of social and political changes over the centuries in order to make a reliable estimate of the times when there was no interest in the heavens, or when observations might have been deliberately suppressed. Partly for this reason, there is no immediate chance of getting any guidance to the 11-year cycle, if it existed, from the observations of 2,000 years ago. But there is very clear evidence of gaps in the record 100 to 200 years long, when very few sunspots were recorded even though there is no political reason why the evidence should be missing. These gaps, shown in Figure 17 which is based on Clark's work, fit in very nicely with the minima found by Eddy in his carbon-14 analysis. The Oriental records don't go back as far as the carbon-14 "record," and they are more patchy. But they confirm the accuracy of the carbon-14 method back to the Medieval Minimum of A.D. 640–710, a full thousand years before the start of the direct record of telescopic observations made in modern times. When all the evidence points the same way in the centuries where it overlaps, we can have a great deal

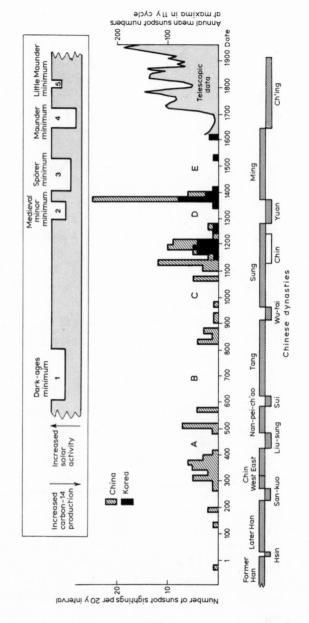

126

*Figure 17*   John Eddy's estimates of solar activity changes, determined from the carbon-14 record, are compared here with the sunspot sightings of Chinese and Korean astronomers. A major feature before the Maunder Minimum is now called the Spörer minimum; in addition to the major features of Table 1, this plot shows (at the top) two small dips in solar activity, the Medieval Minor Minimum around A.D. 1300, and the Little Maunder Minimum of the early nineteenth century. David Clark, who prepared the figure on which this illustration is based, has assessed the gaps in the Oriental record (A–E, middle plot) in terms of the political events of the various dynasties. Gap A occurred at a time of political upheaval in China, when many records were lost, so it may not represent a real decrease in solar activity. Gap B occurred after reunification, during a period when other records (including calendars) are very complete, and is probably a real change in solar activity. Gap C came at a time of low standards in astronomical work, just before the great reforms of the Sung Dynasty, and probably represents the incompetence of the observers rather than changes in the Sun—as the carbon-14 record confirms. The later gaps are confirmed by Korean records, as well as Chinese, so that even though gap D comes at a time when the Mongols were overrunning China we can be sure that it is real, while gap E closely corresponds with the Spörer and Maunder Minima known so well from other studies.

of confidence in believing the one record, the carbon-14 measurements, that goes back furthest into the past. But there is still more to the story of ancient Oriental astronomical observations of the changing Sun and its influence on the Earth.

## LINKING SUN AND EARTH

One astronomer, at least, has been bold enough to try to test for the presence of the 11-year cycle of activity over the entire period covered by the Chinese records. Dr. A. Wittman, of Göttingen University in Germany, has used the reasonable assumption that although it may be just about impossible to use these fragmentary records to search for a periodic variation if you don't know the "answer" in advance, it is a different kettle of fish if we know when the peak years of sunspot activity "ought" to have occurred in the past, and use the records to test whether more sunspots really were observed in those years than at other times. Again, you have to take account of political factors, and there is the risk of a circular argument since, of course, we know the period we hope to find. But Dr. Wittman finds that, putting the most modest possible interpretation on his analysis, fifty maximum sunspot years can be identified between 500 B.C. and A.D. 1600, when the telescopes took over. Of course there are gaps—centuries at a time when no peaks can be identified, just like the end of the seventeenth century. Some of these gaps are due to the Sun being quiet, some to lack of observations. But the key discovery is that all the sunspot peak years occur close to the dates expected simply by running the 11-year cycle backward into the past. As Dr. Wittman puts it, "It is highly likely that the sunspot cycle persisted without interruption throughout this time span. The mean period is equal to 11.135 years."

Today, with the evidence of Robert Dicke's work on the solar chronometer, we would rephrase this to say that throughout the past 2,500 years it seems that the clock inside the Sun has been running with a steady rhythm of 11 years. The sunspots themselves, when they are present, and any changes in solar activity or the solar "constant" which affect the Earth, are just the outward signs of more deep-seated changes in the Sun.

In the late 1970s, however, a new voice began to be heard again in scientific circles after years of silence during the upheavals of the previous years—the voice of China itself. With communication with the West reopened, news began to emerge of studies made by Chinese scientists in many areas, and one of the first to be reported was their latest work on the ancient records of solar activity and effects on the Earth. These reports appeared in journals such as *Acta Astronomica Sinica,* and still avoiding the "cult of the individual," the authorship of teams was identified as "The Ancient Sunspot Records Research

Group" and the like.* Their analysis goes back to 43 B.C., providing a welcome independent check of the historical study made by Clark and others in the West. Like their Western colleagues, the modern Chinese find evidence in their records of Eddy's "Medieval Maximum" of A.D. 1100–1300, and gaps corresponding to the Spörer and the Maunder Minima. But the "Ancient Sunspot" group have gone further by devoting most of their analysis to an attempt to find other periodic variations in sunspot activity, way beyond the basic 11- and 22-year cycles.

Their results are rather unselective, by the standards of present-day Western astronomy, and the team seems eager to seize on any trace of a cycle and claim it is real, trying to find a physical explanation. In the West, astronomers generally have to be hit over the head with a weight of evidence before they believe any of it, as poor Maunder himself found with the discovery of the seventeenth-century minimum. The best path, surely, lies somewhere between these extremes, so perhaps the reopening of scientific links between East and West will benefit both parties! Meanwhile, though, it is worth noting that the Chinese do seem to be persuaded that some influence of the giant planets of the Solar System affects sunspot activity.

Even if you prefer to file this idea under science fiction for the time being, there is no denying the opposite interaction, that changes in the activity of the Sun do affect the planets, and in particular that changes in solar activity affect the Earth. The very strong influence of the Sun on the Earth is shown most simply in new data from another Chinese group, reported in the same scientific journal as the "Ancient Sunspot" analysis.

This team has looked at changes in solar activity and changes in the rate at which the Earth spins since the early nineteenth

*See my report in *New Scientist* 76 (1977): 703.

129

century. There is no problem here of interpreting the evidence in the light of political history; happily we are back in the realm of modern astronomy with complete records available for all to see. Changes in the rate at which the Earth spins—the length of day—are known simply from timing the interval between each "midnight," when a chosen star is at its highest point in the sky (in principle this is easy; in practice it is a very painstaking operation, but we can leave the details aside). Since 1820, the length of the day has changed over a range of about eight milliseconds, leaving aside seasonal effects linked with changes in the atmospheric circulation and so on. Sometimes the length of day increases for a while, sometimes it decreases, in no easily predictable fashion (although some of the changes are linked with earthquake activity; see Chapter 9). The success of the Chinese team is that they have explained all the changes since 1820, and made a forecast of future changes up to the year 2000, simply by adding together twelve periodic effects, most of them related to known patterns of change in the Solar System, such as changing solar activity or changing alignments of the outer planets—or both.

This is an old idea, sparked off originally by the fact that the sunspot cycle itself is just over 11 years long, while the period of Jupiter in its orbit (the Jovian "year") is 11.86 years. The Chinese also find a lesser periodic ripple of 29.8 years, close to the orbital period of Saturn (29.5 years); a periodic ripple very close indeed to the interval between conjunctions of Jupiter *and* Saturn (alignments of the two planets together occur every 19.9 years, and a ripple exactly this length is found in the Chinese study); and the longest reliable period of all, 179 years, which corresponds to the interval between alignments of all the outer planets in the same configuration.

This last hint is the most tantalizing. Because of their movements at different speeds in their orbits at different distances from the Sun, the pattern made by the planets at any time seems

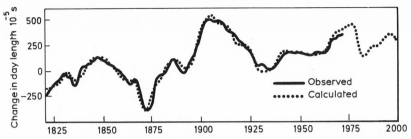

*Figure 18*   Changes in the length of day calculated from the cycles of outside influences, including sunspot activity and changes in the alignments of Jupiter and Saturn (dotted line), exactly match the actual measured changes since the early nineteenth century (solid line).

Changes in the Sun's activity do seem to affect the spin of the Earth—the length of day—and changes in the alignments of large planets, such as Jupiter and to a lesser extent Saturn, do the same trick. This can be understood, perhaps, directly in terms of the gravitational (tidal) forces exerted by the giant planets on the spinning Earth. But there is also a temptation to speculate that, among the complexities being unraveled by this kind of study, the planets themselves—especially Jupiter—may affect the Sun, and the resulting solar disturbances then feed back their effects upon lesser planets such as the Earth.

to be ever-changing. But following the intractable laws of celestial mechanics, including both the influence of the Sun's gravity and little tugs from one another, the planets actually move so that this heavenly clockwork of the Solar System resets itself every 179 years. So the planets are aligned just the same way, as I write this chapter in 1979, as they were in 1800. Even with only 300 years of telescopic sunspot records, some Western astronomers had found a hint of a 179-year cycle in long-term solar activity; now the Chinese, with their study of the spinning Earth, seem to be confirming this.

It takes a bold scientist to include a 179-year periodic effect on the basis of a sample of data only 150 years long, but the

131

"cycle" is needed to make the length-of-day curves fit exactly, and it does tie in with two physically known phenomena, the hint of a long period in sunspot variations and the realignment period of the planets. In both the solar activity variations and the changes in length of day, the cycle is far from dominant— if you like, it is a lesser ripple on the ripple of the 11-year variation. But if there is a clue here that all of the planets acting together do exert an influence on the Sun, no matter how small, then it makes more plausible the suggestion that Jupiter (by far the biggest of the planets, of course, bigger than all the rest put together) is indeed linked with the fundamental 11-year rhythm in some way. It may be hard to see how even a planet as big as Jupiter could do the trick, especially since the sunspot cycle is slightly different from the orbital period of Jupiter, and it might just be a coincidence that the solar clock keeps roughly the same time as Jupiter. But if the other patterns, involving Jupiter-Saturn conjunctions and alignments of all the planets, are also present, then the coincidence is stretched to the breaking point. This isn't yet respectable science, at least in the West. But I'm willing to stick my neck out in accepting the circumstantial evidence as compelling—I believe that in the very near future even Western astronomers will come to accept the influence of the planets on solar activity and will find out how the link works, probably with the aid of the next generation of spaceprobes and their instruments studying the Sun from above the murk of the Earth's atmosphere.

Speculation is allowable in science; indeed, it is often welcome as providing a signpost to the kind of further work that might be fruitful. Even a speculation that later turns out to be wrong is useful if it encourages people to find the right path. It is, however, vital that speculation not be presented as fact or dogmatic belief. The concept of the Sun as a perfect sphere would have been fine as a speculation which encouraged astronomers to look for imperfections, and then to discard the concept of perfection once sunspots were found. It caused trouble as an idea only because it became an article of faith, with established authority refusing to accept the new evidence. So I should be careful here to draw the line between established fact and speculative ideas, not only so that the respectable works don't get tarred with the brush of speculation, but also so that the vaguer ideas don't achieve an aura of belief that they don't yet deserve.

So—the work of Spörer and Maunder decades ago, backed up by the studies of John Eddy, David Clark, and Chinese astronomers in the 1970s, shows beyond any doubt that the Sun is a variable star in a timescale of decades, centuries, and millennia, with the visible sunspots best seen as a surface rash whose comings and goings hint at processes deep in the solar interior. It is *possible,* but by no means proven, that the development of this rash is linked in some way with disturbances from outside, dominated by the (presumably gravitational) influence of Jupiter. And it is *certain* that changes in the level of solar activity influence our planet, the Earth, in such a deep, fundamental way as changing the whole spin of our planet (by a few milliseconds). How else do the solar flickers affect the Earth—and can we make any forecasts of future terrestrial changes linked with those flickers?

# CHAPTER 9

# The Fatal Flickers

NOW COMES THE CRUNCH. WE HAVE SEEN HOW IMPORTANT the Sun and its relationship with our planet are for life on Earth in the broad sense, and it may be that some solar change, caused by changes in the nearby interstellar environment, produced the pattern of ice ages and "interglacials" out of which mankind's civilization has sprung. All that, however, is remote from the here and now, and interesting only in an abstract way. But those flickers of the inconstant Sun described in the last chapter bring us right into the realm of immediate interest. For, quite apart from the climatic influences of our changing Sun, there are short, sharp changes which can spread their fatal destructive influences across space to our planet.

The background against which we must look at such short-term disturbances is amply covered by climatic and solar history since the most recent ice age, the past 10,000 years or so. This period, while only a tiny fraction of the Sun's life, represents the entire history of modern man. Accepting that there are changes in the radiation put out by the Sun, John Eddy points out:

[when] the flow of radiation through the outer solar atmosphere [is] perfectly constant, we might expect a sunspot cycle whose peaks were almost uniform in amplitude. If the flow of radiation were slowly increased, we should expect an overall enhancement of sunspot production . . . in the run of heights of the 11-year period. If the flow of radiation were slightly reduced, the peaks of the cycle would be depressed. And if the radiation fell below some critical level, perhaps only a drop of 1 percent or less, the amplitude of the cycle might be damped so much that the cycle would shut down, or appear to shut down, as during the Maunder Minimum.*

The height of a sunspot peak, or a run of peaks, it seems, tells us how the underlying heat production of the Sun is varying. Putting some numbers in, the peak value of what used to be called the solar "constant" (now the solar parameter) occurs at 1.94 calories per square centimeter per second reaching the Earth's surface, for cycles peaking around 80–100 in sunspot numbers. Some observations suggest that the solar output is a little less than this peak, not only for lower sunspot numbers but also for very strong cycles, peaking with sunspot numbers over 100; but so few cycles of such strength have been observed that this suggestion has to be taken with a large pinch of salt. The *reliable* evidence is that up to a sunspot number of 100 a bigger peak in the cycle means more solar output, with up to a 1 percent warming as far as the Earth is concerned. And a fluctuation of 1 percent or so, while still much less than we can measure in terms of solar changes over several decades, is ample

*Quote from Eddy's talk at the International Symposium on Solar Terrestrial Physics, held in Boulder, Colorado, in June 1976. The *Proceedings,* edited by D. J. Williams, were subsequently published by the American Geophysical Union.

to explain all the climatic fluctuations of the past 10,000 years.

When graphs of changing temperature on Earth over those ten millennia are compared with the changing level of solar activity, shown by carbon-14 changes, the fit is so good that it has been described as "like a key in a lock."* Whenever the solar activity dips, the climate turns cold and glaciers advance; whenever there is a prolonged period of high solar activity, glaciers retreat as the Earth warms up. The range is only a couple of degrees in temperature, comfortably fitting a range of solar variations of 1 percent up or down. These are long-term changes which we can no more detect directly from available records of solar studies than we can from studies of other G type stars.

All of this makes it possible to forecast future weather trends, if you accept the reality of the various cycles found in the patterns of solar variation, up to and including the 179-year cycle. Just such a forecast has been made by Professor Hurd Willett, of the Massachusetts Institute of Technology, who expects temperatures to fall over the next twenty-five years as solar activity declines, with a short warm spell in the early twenty-first century being rapidly followed by a return to colder conditions.† The next period of warmth comparable to the decades from 1930 to 1960, the decades of high solar activity, is not due until the years 2110–2140—and that is far enough ahead for anyone living now to be worried about! But there is plenty to be worried about in the *immediate* future, as we look at the most erratic flickers of the Sun, and to puzzle over some of those hints of a link between solar changes and influences from the giant planets.

*Eddy, *op. cit.* The temperature changes of past millennia are inferred from a variety of geological and other techniques; see my book *Climatic Change.*
†For details of Willett's forecast, see bibliography and the discussion in *What's Wrong with Our Weather?*

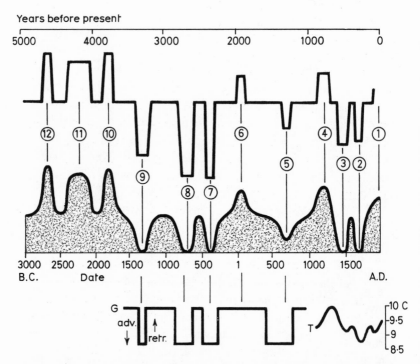

*Figure 19* Like "the fit of a key in a lock," this figure (based on data provided by John Eddy) shows the match between changes in the production of atmospheric carbon-14 (top) and the advance and retreat of glaciers on Earth. The middle, shaded, curve is a smoothed-out version of the carbon-14 variation, representing the changing level of solar activity, with twelve important peaks or dips in the past 5,000 years. For the past thousand years or so, the pattern of glacier movements has been replaced by a temperature curve inferred from various historical data. The message is clear—when the Sun is less active the Earth cools. (Note that the top curve has been *reversed* for this comparison; more carbon-14 in the Earth's atmosphere is associated with a decline in solar activity.)

## CHANGING SPINS—OF SUN AND EARTH

One of the latest puzzles for solar scientists, now increasingly concerned about the evidence of solar variability, has come from studies of astronomical drawings made over the years going back to the invention of the telescope, and recording sunspots in the early seventeenth century. These show that the Sun, which today rotates once every twenty-eight days, began to spin faster in the years before the Little Ice Age. The situation is complicated because the Sun, being a fluid body, doesn't rotate all at the same rate—the "twenty-eight days" is an average, with the high latitudes taking thirty-four days to turn around once and the equatorial region zipping round every twenty-five days or so at present. But, according to several independent assessments of the records (reviewed in an admirably concise "popularization" by Walter Sullivan in the *New York Times*), the rotation of the equator, carrying sunspots around with it, increased by at least 5 percent as the Little Ice Age approached. Why and how this happened nobody yet knows—but it does, perhaps, provide us with a possible early warning system to detect the next Little Ice Age. As a result, astronomers are now keenly watching the Sun's equatorial rotation rate. For now, let's just take this as further evidence of solar variability and imperfection.

The subject of rotation, however, is still a hot topic in discussion of solar-terrestrial links. The Earth itself rotates, as we all know, but as many people do *not* know, the Earth's rotation too is irregular. Only on a small scale—the length of the day varies by a few milliseconds, not much compared with twenty-four hours or so, but impressive behavior for a planet weighing in at $6 \times 10^{24}$ kilograms. So what causes the Earth itself to shiver on its axis?

One certain cause is earthquake activity. When the whole of the Earth is subject to a fit of crustal shudders, as earthquakes

139

and volcanoes burst into life, there are measurable changes in the length of day. This happened, for example, between 1897 and 1914, when there were seventy-one earthquakes of magnitude 8, each roughly as destructive as the famous San Francisco earthquake of 1906. At the same time, there were many volcanic eruptions, and the Earth's spin, indicated by astronomical measurements, slowed down at a rate never before measured. Another cause, regular and predictable this time, is the seasonal movement of great masses of air around the globe, changing the balance of the spinning Earth and altering the length of a day by about 0.0025 seconds and back over a year. And all these alarms and excursions take place against a steady slowing down, at a rate of 0.0016 seconds added to the length of the day each century, caused by lunar and solar tidal effects. But all these reasonably well understood and explained effects don't account for all the changes in the Earth's spin over the years. Especially since the advent of atomic clocks, which have given us unprecedentedly accurate measurements of time since the 1950s, it has become clear that there are other, irregular fluctuations in the length of day still to be explained—and where better to find an explanation than in the irregularly changing Sun?

Several studies have, indeed, shown that there is a very clear link between variations in the overall activity of the Sun, revealed by the changing sunspot number, and the rate of rotation of the Earth. Of course, the effects are small, but they are nonetheless real. And if the "ordinary" changes in solar activity over the 11- and 22-year cycles can measurably affect the spin of the Earth, what about unusually large bursts of flaring activity on the Sun?

It isn't difficult to find an explanation for this solar-terrestrial link, now that we know, thanks to spaceprobes, about the "wind" of charged particles which blows outward from the Sun, stronger and more gusty at sunspot maximum. These particles can affect the atmosphere of the Earth, encouraging low pressure weather systems ("cyclonic disturbances") to develop at high latitudes and thereby altering the spin in much the same way that seasonal changes in atmospheric circulation do. From time to time (usually but not always at sunspot maximum), the Sun produces particularly large flares, and these, with their effects on the solar wind, are enough on their own to produce a measurable shaking of our planet.

Ironically the effect of a giant flare on the changing length of day was noticed and reported by French astronomer Dr. A. Danjon as long ago as 1962, before we knew much about the solar wind at all. Indeed, at that time Danjon's observations were so surprising and inexplicable that they were largely ignored. As later studies in the 1960s showed the reality of the solar activity–length of day link, Danjon's discovery should have received more attention, but it seems to have got lost among the flood of scientific papers appearing all the time. In the early 1970s, while I was working on some related problems with Dr. Stephen Plagemann, then of NASA, I came across Danjon's papers. We would have liked to find a way to test his observations that one large solar flare can slow the Earth down measurably, but with the next sunspot peak not due until the 1980s it seemed we were in for a long wait. Then, in 1972, quite unexpectedly and at a time of relatively low sunspot activity, the Sun coughed up the biggest flare ever recorded—further proof that there are still mysteries to be solved in solar physics! The obvious thing to do was look for an effect of this solar storm on the changing length of day, and Dr. Plagemann and I analyzed data from the US Naval Observatory in Washington which monitors the spin of the Earth using astronomical observations and atomic clocks.

To our delight, we found exactly the effect needed to fit in with Danjon's observations of a decade before—a small but sudden increase in the length of day after the flare, followed by a gradual return to "normal" as the spinning Earth settled down. Curiously, though, when other astronomers tried to check this out using averaged-out data supplied by the Bureau International de l'Heure in Paris, they couldn't find the "glitch" which is so obvious in the Washington data. In other words, from their point of view the little glitch we found was just a coincidence—an erratic jump in the length of day, which just happened to have come exactly in the right place to match the solar flare and which, coincidentally, is the biggest such jump in the record of the changing length of day for three months on either side of the date of the solar outburst.

Which interpretation you prefer depends on how far you are willing to stretch the long arm of coincidence. I still believe the solar storm of August 1972 shook the Earth, and now wait for the next peak of solar activity to provide further confirmation that such outbursts do affect our planet in this way.*

Solar flares, however, are certainly of great importance, regardless of how strong the link with sudden changes in the Earth's rotation turns out to be—and I still believe it is strong. It is the buildup of solar activity toward the next peak that has caused the eighty-five-ton Skylab to fall out of the sky years sooner than NASA expected, as the effect of a stronger solar wind heated the upper atmosphere, changed its density, and increased drag on the massive satellite. The August 1972 event itself produced a decrease of more than 10 percent in the ozone concentration of the stratosphere, as protons from the flare plowed into the atmosphere and disrupted the chemical balance high above the ground. Such an event can be coped with—but some thoughtful astronomers now wonder whether still bigger flares could do long-term damage to the environment of the Earth, especially if they happen at a time when the Earth's shielding magnetic field is weak. A giant solar flare at the right

*For more details of this debate, see *The Jupiter Effect.*

142

(or wrong!) time could all but remove the ozone layer which protects us from ultraviolet radiation, with the result that the land surface could be almost sterilized, with large plants and animals being wiped out, before the natural chemical balance is restored. Can this, some astronomers wonder, explain the sudden end of the era of the dinosaurs, about 65 million years ago?

And although there seems to be very little proof that individual solar flares and sunspots affect temperature on Earth, there is clear evidence that outbursts from the Sun do stimulate the development of high-latitude storm systems, as I have described in *What's Wrong with Our Weather?* Even in this day and age, we are still very dependent on radio communications, which can be disrupted by charged particles from a gusty solar wind interacting with the Earth's magnetic field. And, of course, as space travel into Earth orbit is becoming routine, it is increasingly important to avoid exposing astronauts and scientists to any more of this radiation than necessary.

With all these undoubted effects, there is indeed the possibility that the effect of flares on the rotation of the Earth is real. And since we know that earthquakes can change the length of day, is it possible that sudden changes in the length of day—a jolt to the spinning Earth—can trigger earthquakes? How nice it would be if we could only find some way to forecast not just the year of the next sunspot maximum, and its size, but also the actual dates on which particularly large flares on the Sun can be expected. At the very least, identifying the year of solar maximum would tell us when the averaged solar influence on the length of day is producing maximum braking effect—and that could neatly explain the long known "coincidence" that outbreaks of earthquake activity seem to be in step with the solar rhythm.

Well, there's some good news and some bad news. The good news is that there does exist a reliable, proven method for forecasting changes in solar activity, a technique of benefit to the radio industry, space agencies, weather forecasters, and now, perhaps, to earthquake prediction. The bad news is that the technique depends on analysis of the positions of the planets relative to the Sun—something so close to astrology that establishment science refuses to touch it with a barge pole.

## THE ASTROLOGICAL CONNECTION

The word "astrology" creates such powerful emotions, from both sides of a very tall dividing fence, that it is important to make it clear from the outset what I am talking about. I don't believe that the "horoscopes" in daily papers or monthly magazines are anything more than a harmless bit of fun, and I am very doubtful about the claims that our personalities are molded by the influence of the planets (I say "very doubtful" because every scientific study I know of designed to test the idea has failed to resolve the issue one way or the other). In any case, that is a subject for the psychologists to tackle. But, as an astronomer with an interest in Earth sciences, it clearly is appropriate for me to consider whether there is a real *physical* influence of the Sun and planets on physical processes operating on Earth. This is no more "astrology" in the daily horoscope sense than it is "astrology" to point out that the Moon affects the weather (which it does, through tidal effects). And there is no doubt at all that the alignments of the planets do play a part in determining the occurrence of large flares on the Sun, while perhaps modulating the entire solar cycle of activity.

That said, anyone who ventures from the cozy security of established science to look on the other side of the fence finds quite a mess. Because *some* things regarded as "unclean" because of the taint of astrology are good, many people jump to the conclusion that *everything* dismissed by conventional science must contain more than a grain of truth! This is no more realistic than assuming that it is all rubbish. So I am really attempting to pick my way across a mine field in attempting to summarize some of the minority of unconventional ideas which do seem relevant to the puzzle of solar activity and its variations, and which certainly merit more serious attention than they usually get.

Sometimes, though, scientists stumble on the significance of planetary alignments without realizing that this is a taboo topic. The classic example is Dr. John Nelson, a radio communications engineer who was working with RCA in the mid-1940s, and had the task of finding some way to forecast when outbursts of solar activity were likely to occur, disrupting communications by the influence of the solar wind on the ionosphere of our planet. In his own words, he was asked to forecast "radio weather"; and since the radio weather depends on solar activity, that amounted to finding a way to predict outbursts on the Sun, and the peaks of sunspot production. Nelson soon found that some sunspots are more likely than others to be associated with flares that disturb the radio weather, and with this aid was soon achieving about 70 percent accuracy in his forecasts for RCA staff scattered around the world. Trying to improve this accuracy, he followed up the many published suggestions that the planets influence the Sun—suggestions which a professional astronomer at the time would have dismissed as unworthy of serious attention since "of course" the Sun, with 99 percent of the mass of the whole Solar System, couldn't be disturbed by the tiny gravitational forces of the planets. Without this dogma to blinker him, Nelson went right ahead comparing the positions of the planets on occasions when the Sun produced particularly nasty flares. He found that when planets such as Venus, the Earth, Mars, Jupiter, and Saturn were arranged so that they made right-angle and 180° (straight line) alignments with the Sun, then strong flares and disturbed radio weather could be expected. When the planets made 60° or 120° angles, conditions were likely to be good.* Accepting this at face value, Nelson boosted the accuracy of his forecast technique to 85 percent, and was now able to forecast both good and bad radio weather, using the two significant kinds of alignment. Although RCA announced this in 1951, and continued to use the "astrological" technique successfully throughout the subsequent decades, it made no splash in scientific circles, where it has remained virtually ignored to this day.

Why should this be? Chiefly prejudice—for if you talk to an astrologer, one of the first things you learn is that angles of 60° and 120° ("trine") are "good things," while right angles ("square") and 180° alignments ("conjunction") are basically "bad things" to find in a horoscope. Johannes Kepler would have accepted Nelson's results without question; but mid-twentieth-century astronomy "knew" that everything in astrology was wrong, so it "knew" that Nelson's work, with its astrological overtones, could be ignored. Even so, RCA, hardly a gullible organization, continued to use this "impossible" technique for one simple reason—it worked and saved them money.

Now let's stand back a bit and look with open minds at how the effect might work. First, let's look at that point about the "impossibility" of the little planets disturbing the equilibrium of the Sun, sitting like a rock at the center of the Solar System. Well, hold on. Actually that's wrong for a start. The Sun is *not* at the center of the Solar System!

The planets may be small, but they are significant in setting

---

*The use of the term alignment needs some explanation. In an earlier book, *The Jupiter Effect,* I referred to an "alignment" of the giant planets, and many readers took this to mean that they would be "lined up" in a straight line. *Any* pattern of the planets is an "alignment," borrowing the term from astrology, including patterns that make right angles or 60° angles centered on the Sun. More of this later!

146

up the balance of the Solar System, and the center of the system is the center of mass, averaging over the positions and masses of all the planets *and* the Sun. Jupiter is the biggest planet and dominates the others in establishing the balance with the Sun, but all play their part. Just as a very small boy on one side of a seesaw can balance an adult on the other side provided the adult sits near the point of balance (the center of mass then coincides with the point of balance), so the massive Sun close to the center of mass of the Solar System "balances" Jupiter, a long way away from the center of mass on the other side. But all the planets are orbiting around the center of the Solar System (*not* around the Sun!), and the Sun too is in orbit about the common center of mass. Because the planets move at different speeds in their orbits, the Sun is tugged a little bit this way, then a little bit that way, while mainly following the motion of Jupiter, of course. From the point of view of the Sun, the center of mass seems to follow a looping orbit, sometimes below the surface of the Sun, sometimes outside the Sun altogether. But that is a pretty crazy way to look at it.

From the point of view of an imaginary observer sitting at the center of mass—the true central pivot of the whole Solar System—the planets wheel about in their orbits, much as we are used to thinking of their movement through space. But now we see the Sun itself moving in its own looping orbit, counterbalancing the constantly changing average effect of all the planets. The basic beat of the Sun's orbit is its reflection of Jupiter's movement—and remember Jupiter takes just over 11 years to complete one orbit of the Solar System. The wiggles which disturb this basic pattern depend on how the other planets move, and remember that the whole pattern of Solar System clockwork repeats after 179 years. So here, immediately, we find that the Sun itself is swinging around in a complex pattern with a basic 11-year rhythm and an overall cycle of 179 years. In physical terms, it's like jiggling a bucket of water on the end of a piece of elastic with periodic jiggles taking 11 years and 179 years. No physicist would be surprised to find waves in such a bucket with exactly these periods—so why should we be surprised to find these cycles, characteristic of the motion of the whole Sun, in the rhythms of solar activity?

We used to think of the Sun as the center of the Universe (once we realized that the Earth wasn't at the center of the Universe!), and even though this position of eminence has been displaced we still find it hard to believe the Sun is not the great central pivot of the Solar System. This conceptual difficulty alone seems to have prevented progress in understanding the rhythms of solar activity. Once we appreciate that the Sun is not even the center of the Solar System, but must move about the true center in obedience to the same physical laws that govern the movements of the planets, it becomes much easier to accept the reality of findings like those of John Nelson.

We still don't know *why* some planetary alignments encourage solar activity and others seem to suppress it, but that is hardly the point. The job of theorists is to explain the observations; it isn't up to the observers to try to fit observations to an outmoded theory! We can even, whisper it low, explain why astrology got the idea that squares are bad and trines are good. We now know that these alignments do affect the Sun; we also know that solar flares disturb the environment of the Earth, through influences on weather and the electromagnetic field. Over many generations, astrologers may well have noticed some simple correlation between moods and the alignments mentioned above—perhaps people are more miserable when the Sun is active, the sky is cloudy, and the ionization of the air changes. They could easily do this without realizing that the Sun is a necessary step in the chain, or even without appreciating that the Sun varies at all. But that speculation I leave for modern astrologers themselves to pursue.

The motion of the Sun relative to the center of mass of the Solar System has received some attention from scientists, notably Dr. Paul Jose, in a study for the US Air Force, in which he predicted the peaks and troughs of the sunspot cycles to be expected in the decades after the 1960s. He was right in predicting the maximum in 1969 and minimum in 1977; the next forecast is for a maximum in the early 1980s. In another report, this time for NASA, H. Prescott Sleeper used analysis of planetary alignments to make similar predictions in 1972, building to some extent on an earlier NASA report prepared by Dr. J. B. Blizard, of the University of Denver. The sponsors of these reports are significant, for like RCA both the US Air Force and NASA have good reason to want to know when solar flares are coming—they have satellites in orbit that might be damaged, quite apart from the desirability of scheduling manned flights to avoid the worst dangers.

So a pattern emerges—people who really need to know what the Sun is doing use the planetary alignment predictions. Just the same pattern occurs with dowsers—many scientists dismiss dowsing as "obviously" impossible. But many commercial organizations employ dowsers to find buried pipes, water, or even minerals, and they do it for one good reason—the technique works and it is cost effective.

## FORECASTS—FOR WHAT THEY ARE WORTH

Using this philosophy—that a tool can be useful even if we don't know how it works (how many drivers understand exactly how their cars tick?)—we can pick out a few of the forecasts related to coming solar activity which may affect us here on Earth. First, perhaps, my own forecast, made some years ago in *The Jupiter Effect.* This rests on two basic bricks, that the southern part of California, near Los Angeles, is long overdue for a major earthquake (by which I mean one as big as the San Francisco earthquake of 1906), and that peak levels of solar activity are associated with outbursts of earthquake activity, through the changes in spin already mentioned which result from a strong and gusty solar wind. When the next peak of solar activity arrives in the early 1980s, the whole Earth will be shaken by a small amount, triggering regions prone to earthquakes where strain has been accumulating for some time. This is *not* an "end of the world" prediction, just a forecast of more than average earthquake activity. Since the southern San Andreas fault has regions where strain has been building for more than one hundred years, it seems particularly likely to go if there is any increase in the level of earthquake activity and resultant release of tension around the globe.

When Steve Plagemann and I first made this forecast, we wanted to try to pin down the exact year of the next sunspot maximum, and plumped for 1982 because that is going to be the time of the next "grand alignment" of all the planets in the outer Solar System, a time when all the giant planets are on the same side of the Sun, gathered within a very narrow angle. This, we thought, ought to be a strong disturbing influence for the Sun, marking the time when the center of the Sun is farthest away from the center of mass of the Solar System, as the Sun pulls out on the other side to "balance" the added effect of all the planets essentially pulling together. But this alignment, although rare (and not to be repeated for 179 years), is not a "straight line" lineup of all the planets; rather, they tuck inside an angle of about $60°$ as viewed from the Sun. We've been taken to task, quite rightly, for not making this clear enough in our original book, but I hope the record is now straight, even if the alignment isn't! More importantly, though, we've also been taken to task by astrologers who point out that the $60°$ alignment is beneficial, and that therefore a major earthquake disaster shouldn't happen at that time. Leaving aside the astrology,

150

John Nelson's method would also suggest that the 60° alignment is not a time to expect great disturbances in terms of solar flares and their effects on Earth.

As a physicist, and depending on the idea of motion relative to the center of mass as a key to the understanding of solar activity, I'm still suspicious of this unusual pattern of the planets. But let me offer for your consideration another alignment which is part of the slow gathering of the planets toward this rendezvous. One of the astrologers I met as a result of the publication of *The Jupiter Effect* prepared for me the chart (Figure 20) shown below, for midnight on December 1, 1980, with all the planets within a right-angle alignment. According to Nelson, that might well be a trigger for solar activity producing bad radio weather; according to my astrologer friend, the position of the Moon is also significant for any earthquake

*Figure 20* An astrological chart showing the positions of the planets, Sun, and Moon at midnight GMT, December 1, 1980. Astrologers have long thought such alignments herald dramatic events on Earth; there is now at least some scientific evidence to support the idea, through a link between planetary alignments and solar activity. (Based on data provided by P. A. Bennett.)

151

trigger effect, and on this basis she says the time to watch for is one hour earlier than the time this chart corresponds to: 11 P.M. GMT on December 1, 1980. I don't quite believe it—and yet, I shall be very careful to keep away from earthquake prone regions early in December!

The solar effect, though, may be more significant. This alignment really should correspond with something interesting on the Sun, and that is what this book is all about. (And, for anyone who has been wondering, the great solar flare of August 1972, and the earlier event in 1960 whose effects were monitored by Dr. Danjon, both corresponded to significant Nelson alignments.) There is certainly enough evidence here, whatever happens in December 1980, or in 1982, that the ideas developed by Nelson, Sleeper, Jose, and Blizard ought to form the foundation of future efforts to understand the workings of the various solar cycles of activity.

Joseph Goodavage, a writer and "astrometeorologist," is one of those people who discovered a nugget of truth in "astrology" and, it seems to me, has thereby been encouraged to take at face value a lot of fool's gold as well. But since one of my own main interests is the weather, and Joe Goodavage has stuck his neck out in print by making "astrometeorological" forecasts for the decades ahead, it seems worth searching through the fool's gold to find any more nuggets that might be lying around. The story of how Goodavage got sucked into this kind of study is interesting in itself—as a science writer, he set out to debunk astrology and found instead that he had been convinced that it stands on a secure foundation, so ever since he has devoted his efforts to persuading other scientifically literate people that there is something here that merits investigation.

The weather forecasts are particularly intriguing. Goodavage has had a long association with the magazine *Analog,* which publishes both science fiction and science fact. In his recent book *Our Threatened Planet* he describes a test carried out in

152

the 1960s by the then editor, John W. Campbell, in which astrometeorological forecasts were published alongside those of the US Weather Bureau and a random forecast generated by a computer. At the end of the experiment, the system based on planetary alignments was rated 94 percent accurate, the Weather Bureau forecast about 50:50, and the random forecast actually came out better than the Weather Bureau! So why didn't the Weather Bureau take up astrometeorology? A clue may be found in the attitude of a physics professor who had once taught John Campbell himself, in an anecdote of Campbell's which Goodavage reports. Not long after Campbell obtained his physics degree, he was sitting on his front porch when he saw a ball of lightning form and bounce across a field before exploding and destroying a small oak tree. When he next visited MIT, he called on his former teacher to report the phenomenon; the professor assured him that ball lightning did not exist, and that "no competent observer had ever reported the phenomenon."

Here is catch 22—no *competent* observer had ever seen ball lightning, and anyone who claimed to see it must be incompetent, fresh physics degree or not. The story goes back many years, and today physicists do accept that ball lightning exists, although we still don't understand it. So will it take thirty years for astrometeorology to become respectable? Probably a lot less, if Goodavage's latest published forecasts turn out to be accurate.

The picture he paints in *Our Threatened Planet* is one of almost unrelieved gloom:

[A] bitterly cold winter of 1981, particularly in the northeastern and northwestern and border states of the USA, throughout the Canadian northwest, and especially in Europe and the Soviet Union. The Deep Freeze of 1981 will last from the time of the winter solstice of December

153

21, 1980, through January, February, and March of 1981, more intensely so during Saturn's stationary position centering on January 25, 1981. This promises to be the most bitterly cold week of the most bitterly cold winter within living memory.

Who said astrologers always couched their forecasts in ambiguous terms open to many possible interpretations! Nothing could be more precise and dramatic than this, and Goodavage's forecasts for the 1980s tell a similar tale throughout. It is hard to take at face value. But then, remember that we are entering a period of low solar activity according to all the best evidence, and can expect a retreat from the warmth of the mid-twentieth century and a return to Little Ice Age conditions. Goodavage's graphic detail is the spelling out of the implications of broader trends noticed by many other people. The difference is, one mild winter in 1981 wrecks his forecast, while the more general patterns studied by, say, Professor Hurd Willett, can stand a couple of mild winters in the early 1980s as long as the slow downturn of climate continues over the next couple of decades.

Important though all this might be for mankind, though, what happens next year is really too soon to tell us anything about the current disturbances inside the Sun. The valuable contribution of all this evidence to my basic theme is that it ties in so well with the idea that the Sun today is in a temporary phase of instability—it is off color, sickly, or however you like to think of it. Because the delicate balance inside has been disturbed—which we know, for example, from the solar neutrino studies—we cannot assume that the convective regions are in firm control of the situation. While the Sun is struggling to restore its natural balance and return to conditions that are normal on a timescale of millions of years, it is being tugged about by the planets as it orbits the center of mass of the Solar System, and this disturbing influence may be enough to set the rhythm of internal and surface fluctuations on the very rapid timescale of decades. Quite possibly, when all is well with the Sun, when nuclear burning proceeds at the center and convection is going on in orderly fashion, then the influence of the planets is indeed negligible. Perhaps the dinosaurs, had they been intelligent, would never have developed any astrological beliefs. But here and now the Sun is *not* normal and can be influenced even by small outside disturbances, changing in its turn to produce effects which then spread out to the planets of the Solar System and ourselves.

We've had some fun in this chapter, along with the hard facts, by looking at some ideas which mainstream science might dismiss.* But the underlying theme remains, the sobering thought that the Sun today is not a normal star, that something has disturbed it during the past few million years. Was it the passage through the dense dust clouds edging the Orion spiral arm? Is it "just one of those things" that happens to G type stars from time to time, that they go "off the boil" for a few million years? Or is it possible that the instability and inconstancy of our Sun today is linked to far more sweeping cosmic influences, changes that affect our whole Milky Way Galaxy and all its thousands of millions of stars? Most astronomers think of the Galaxy as a stable place—but most astronomers thought of the Sun as "perfect, and if not perfect, regular" until quite recently. The whole concept of permanence, regularity, and perfection looks increasingly poor as a description of the Universe—so it should be no surprise that some astronomers now question the

---

*At least in the West. Soviet science seems more receptive to all these ideas (including dowsing); see M. N. Gnevyshev and A.I.OL' (editors), *Effects of Solar Activity on the Earth's Atmosphere and Biosphere.*

idea of a constant galaxy, with implications that may transform not just our understanding of the Universe, but our interpretation of the geological record of past catastrophes here on Earth, as well as offering a new insight into the temporary flickering of the Sun.

CHAPTER 10

# Inconstant Galaxy?

NOT LONG AGO IT SEEMED THAT, JUST AS ASTRONOMERS believed they had a reliable picture of a constant Sun, so they also had a fairly complete understanding of our Milky Way Galaxy as an orderly system in which stars wheel around the central nucleus in roughly circular orbits, forming the beautiful spiral pattern (etched by the dark lanes of gas and dust between the stars) so characteristic of spiral galaxies. Around this spirally patterned disk, we can distinguish clusters of stars, almost galaxies in miniature, which among them fill a spherical volume of space much greater than the volume of the thin disk. These are the oldest stars in the Galaxy, formed when the Galaxy itself was created, and containing only the original hydrogen and helium of the primordial matter, plus the elements they have built up for themselves in their own interiors. This halo population really is stable and long-lived, or its stars could not have remained in this almost "fossil" state. As far as change in the Galaxy is concerned, we can ignore them. But the story of a "stable" disk population is now beginning to look rather inadequate—that beautiful picture of a few years ago is starting to fray at the edges and come apart at the seams.

157

This transformation is the result of studies of both our own Galaxy and others, two different perspectives which tell the same story. And the story is one of violence, across the Universe. Where once astronomers thought that "quiet" spiral galaxies like our own, or quiet elliptical galaxies, cigar-shaped, full of old stars and with no spiraling disk, were the dominant and typical features of the Universe at large, over the decades since the invention of radio astronomy every new observing technique has shown quiet galaxies to be the exception rather than the rule.

At the greatest extreme of energetic phenomena are the quasars, discovered in the early 1960s. These appear like stars on photographs of the sky, but only because of their great distances. In fact, the furthest known object is a quasar roughly 10,000 million light-years away, blasting away from us at 90 percent of the speed of light in the universal expansion which has proceeded since the Big Bang. To be visible at such enormous distances, these objects must be shining with mind-boggling energy—10,000 times as bright as the whole of our Milky Way Galaxy. Coming down the scale of violence, there are the so-called BL Lac type objects, which can be seen as galaxies rather like our own, but dominated by an intensely bright, quasarlike nucleus. Seyfert and N-galaxies are quieter still, more like galaxies than quasars, but still with strongly energetic central regions; and there is a wealth of radio galaxies, many of which show energetic features at radio frequencies stretching out across the Universe on either side of what seems to ordinary telescopes to be a quiet and ineffectual little galaxy. At other wavelengths, some of these objects have been shown to be intense emitters of X rays, infrared radiation, and so on. And this is the most basic conceptual difference between astronomy now and twenty years ago. Then, it seemed that the Universe was a quiet place where stability ruled; today, the picture is one of violence and change.

## SUPERMASSIVE BLACK HOLES?

Change is inevitable because these brightly shining objects simply cannot keep pouring out energy at such a rate. Nuclear fusion is totally inadequate to account for the outpourings; it must be that matter is being converted to energy, in line with $E = mc^2$, by some even more efficient process, which would destroy all the mass of a quasar in only a thousand million years or so. So what processes are involved, and how long can a quasar—or active galaxy—"live"? Taking the second question first, the most reliable evidence now is that rather than some galaxies being active all the time, *every* galaxy undergoes violent outbursts from time to time. If one spiral in every hundred that we see is a Seyfert, does it make more sense to argue that these 1 percent of all spirals are special in some way, or to accept that all spirals suffer outbursts of Seyfert activity which total 1 percent of their lifetimes (perhaps 1,000 million years of quiet, followed by an outburst lasting 100 million years, and so on)? Already, the evidence suggests that our own Galaxy, surely a typical spiral, suffers repeated violent outbursts somewhere on the scale from a Seyfert to a quasar. How often and how strong? I'll return to these puzzles shortly, but first there's the question of where the energy comes from.

Conventional wisdom today has it that the energy source of quasars, radio galaxies, and all the rest of these phenomena must be the black hole. The only way to get a sizeable fraction of the energy ($mc^2$) locked up in a lump of matter is to drop it down a very deep gravitational "well," into a very massive object beneath. Even the energy of nuclear fusion, the energy of the hydrogen bomb or the Sun and stars, is peanuts compared to this, and calculations show that with a central mass of millions of Suns, each piece of infalling matter could give up 10 or 20 percent of its rest mass as energy. A mass of millions of Suns tucked into a compact volume of space at the heart of

159

a galaxy must, of course, be a black hole, on a much more impressive scale than the little black holes which might be formed out of dying stars. So the conventional picture explains quasars in terms of black holes of perhaps 100 million solar masses, swallowing up from their surrounding galaxies only 1 or 2 solar masses of material each year, but converting 20 percent of the mass of that material into energy, a blaze of radiation that we can detect across 10,000 million light-years of space.

On the face of things, this looks good. But there is still another question—where did the 100 million solar masses of matter come from to make the central black hole? It could be that when a galaxy forms there will be a lot of massive stars in its central regions, where the original gas cloud of the proto-galaxy is most dense. These massive stars could rapidly zip through their life cycles and die, forming black holes which then collide and amalgamate, building up a supermassive black hole nucleus. But all this takes time. Indeed, forming a galaxy out of a cloud of swirling gas takes time too, and most calculations suggest that left to its own devices a cloud of gas in the Universe won't form a galaxy within 20,000 million years— which is embarrassing, since we are fairly happy with the estimates of the age of the Universe as about 20,000 million years, since the Big Bang of creation.

Still, even if you have formed a galaxy, the growth of the supermassive nucleus takes even more time, and the quasar (or Seyfert) activity can only begin once the supermassive core has formed. On this picture, violent galaxies should be old galaxies. But the observations show more quasars a long way away from us than nearby. Because light takes a finite time to reach us, the further away a galaxy is the older is the light we view it by. So when we find a quasar 10,000 million light-years away, we see it as it was 10,000 million years ago, only 10,000 million years after the Big Bang. Most quasars, then, were at their peak of activity much sooner after the Big Bang even than the age of our own Galaxy! This makes the problem of growing the super-mass quickly enough much, much more difficult. Can we resolve this problem and the problem of galaxy formation in one bound?

I think we can, but its only fair to say that many astronomers and cosmologists still don't like the kind of explanation I favor. Quite simply, we can speed up the growth of galaxies by assuming that from the very beginning of the Universe, the Big Bang itself, there were "lumps" of matter which could act as the "seeds" around which galaxies grow. The gravity of these lumps would hold clouds of gas together, allowing stars and galaxies to form in the expanding Universe much more quickly than they could otherwise. And the mass you need to hold together a typical galaxyful of material turns out to be around 100 million solar masses—just the same sort of size we need for the collapsed objects which conventional wisdom expects to find at the centers of active galaxies.*

Whether you accept this resolution of the dilemma or not, there is now very little doubt that there must be supermassive objects at the centers of most, or all, galaxies, however they got there. Some of the evidence can be found close to home, for example in the galaxy Centaurus A, about 16 million light-years away, which has a very large but relatively weak radio source around it. This might plausibly be explained, says cosmologist Martin Rees, of the University of Cambridge, as a powerful radio galaxy or quasar on its last legs, in which case there should still be some evidence of activity at the center of the galaxy, from which an outburst long ago created the surrounding radio "lobes." Sure enough, sensitive modern detectors find at the heart of the galaxy a tiny radio source and an

---

*The problem of galaxy formation is discussed in my book *Galaxy Formation*, and some of these ideas are elaborated on in *White Holes*.

161

X-ray source, flickering on timescales of hours to days as the last dregs of the surrounding material drain into the now quiet black hole, a black hole which in this case must be at least 10 million solar masses.

And even closer to home? Yes indeed, there is something peculiar at the center of our own Galaxy, a compact radio source surrounded by a swirling mass of material. Details of the movements of the surrounding gas clouds (determined by spectroscopic observations in the infrared) suggest that the mass of our very own massive compact object may be only 5 million solar masses, half that of Centaurus A, and not enough for our Galaxy ever to have been a really spectacular radio galaxy or quasar. But still, 5 million Suns make a lot of mass by any normal standard, and that could have been involved in outbursts dramatic enough by human standards—with repercussions for the Sun and Earth.*

## AT THE HEART OF THE MILKY WAY

Now it's time to take up the other thread of the story. Infrared, radio, X rays, and gamma rays are all being used to probe the galactic center, revealing a site of activity scarcely equivalent to a Seyfert galaxy at present, but perhaps marking the power source for occasional Seyfert outbursts. At the heart of the Milky Way, astronomers find streams of gas ejected from the plane of the disk itself, a circular ring of cool clouds right around the nucleus, and outward-moving features both on our

---

*It may be that matter finds it harder to settle in the center of spiral galaxies because of their rotation (angular momentum). So the nonrotating elliptical galaxies could have much greater masses of material falling right into the center to participate in violent activity, which might explain why all the really spectacular energetic galaxies seem to be ellipticals.

162

side of the Galaxy and across on the other side of the nucleus. The pattern of these outward movements, or "expansions," gives us the next piece of the picture.

Naturally enough, just as we once thought of the Earth, and then the Sun, as the center of the Universe, our observations of stellar motions are still often based on what is called the "local standard of rest" or LSR, a system in which the average movement of the Sun and nearby stars is set at zero. In this frame of reference, the Doppler shift in the spectral lines of center of the Galaxy shows a movement of 40 km/sec away from us— in other words, from our point of view the candidate for the nuclear collapsed object is receding from us at this speed. That doesn't make sense, since the center of our Galaxy must be "at rest" in some fundamental way, with the stars wheeling about it in their orbits. But hold on a minute, and look at some of the other velocity measurements.

The "expanding" features show a curious lack of symmetry about the nucleus, at least in velocity terms. These features are clouds of gas and other interstellar material, basically tracing out the shape of the spiral arms. One, between us and the center of the Galaxy, is moving our way at about 55 km/sec; this is called the "three kiloparsec arm" because it is 3 kpc (about ten light-years) from the galactic center. At the same distance on the other side of the nucleus is its counterpart—but that arm is moving away from us at about 135 km/sec, not the 55 km/sec that would correspond to a symmetry of velocities as well as of shape.

Now our Galaxy isn't lopsided, and it would be a remarkable coincidence if these features, moving at different speeds, just happened to be symmetrical about the nucleus at present. The solution is simplicity itself—all we have to do is change our point of view from the LSR to the center of the Galaxy. That 40 km/sec doesn't "really" mean the nucleus is moving away from us, but that we are moving outward at 40 km/sec relative to the nucleus. So we have to compensate for our motion in order to get the true velocities of the expanding features relative to the nucleus, which is what really matters. For the 3 kpc arm, coming toward us, we have to add the velocities, showing the arm to be moving away from the nucleus at 95 km/sec. For the 135 km/sec feature on the other side, moving the opposite way, we have to subtract 40 km/sec, again giving a true velocity relative to the nucleus of 95 km/sec. Symmetry is restored— but only by accepting that the Galaxy as a whole seems to be expanding!

This would be a dramatic discovery if we didn't know about Seyfert galaxies, quasars, and so on. Indeed, most astronomers, brought up in the days when it was thought that stability reigned supreme, still fling up their hands in horror at the thought. But the evidence is indisputable, showing not one but at least two waves of expansion, the 40 km/sec corresponding to star movements in our neighborhood, and the 85 km/sec of the expanding features between us and the nucleus, which are actually catching up with us. Instead of simple circular orbits, stars seem to move in a decidedly noncircular motion, with an outward push added. And this outward wave must in due course be followed by a wobble back toward the center of the Galaxy, or the whole system would have spread out into an ever-thinning disk. So the new pattern we have to explain is one of outward and inward oscillations superimposed on the basic circular motions.

Dr. Victor Clube, of the Scottish Royal Observatory in Edinburgh, has set out over the past few years to explain these noncircular motions and their implications for understanding the structure, history, and future of our Galaxy. Overall, Clube has gathered an impressive weight of evidence supporting the idea that most of the gas and young stars (less than about 500 million years old) in the Galaxy are moving outward with velocities that correlate with distance from the nucleus, with

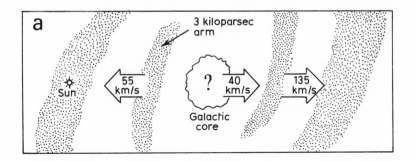

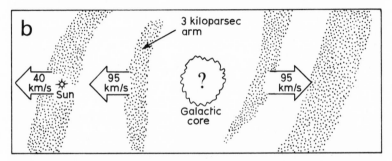

*Figure 21* Two views of our Galaxy: a) the Earth-centered view, in which it is hard to reconcile the physical symmetry of the Galaxy with an apparent asymmetry of motion; b) symmetry restored by measuring velocities relative to the center of the Galaxy.

inner regions moving out more rapidly to overtake the older material, which is further out and now slowing down before it eventually falls back inward. To retain a symmetrical picture, such as we see in most spiral galaxies, Clube needs to invoke repeated outbursts from the nucleus, a Galaxy racked by repetitive upheavals which recur at intervals of one or two hundred million years. Each explosion must temporarily create the pattern of spiral arms that we see now, but in turn each spiral wave

165

will be dissipated as the turmoil works its way out of the system.

Everything seems to hang together, but what makes the Galaxy "breathe" in and out like this? It's at this point that Clube loses the support of establishment astronomy, since he believes that something quite beyond our present understanding of physics may be needed to explain the phenomenon. He suggests that a supermassive concentration of matter like that at the center of our Galaxy, or at the heart of quasars and radio galaxies, does not behave exactly in line with General Relativity as a well-behaved black hole, but that changes in the nature of gravity itself occur at very high densities of matter—changes capable of first squeezing the whole Galaxy into a more compact state and then blasting it outward in a violent rebound, an explosion of colossal dimensions, but no more dramatic than the outburst we see in many active galaxies.

All of this ties in with the idea of peculiar supermassive objects as the "seeds" of galaxies, with the ideas put forward by such respected cosmologists as Sir Fred Hoyle regarding the possible need for a "new physics" to transcend relativity theory just as relativity transcended Newtonian theory, and with other hints that even if pushed to the black-hole limit conventional physics cannot explain all the violence of the Universe (see *White Holes*). These are intriguing questions on a literally cosmic scale, and one day, no doubt, the story will make a fascinating book once someone weaves the strands into a complete new vision of the Universe. It is far too early yet for that to happen; but it is not too early to ponder on how these galactic upheavals would affect the Solar System. Don't forget, however bizarre the new ideas may be, the basis of Clube's work is *observational* evidence for repeated upheavals of the Milky Way system, and we can consider the implications without understanding the cause, just as our ancestors could make plans involving planting of crops and harvesting them without knowing *why* the seasons follow their regular rhythm.

166

## THE IMPACT ON EARTH

The history of the Earth is one of turmoil and change. As I hope I have shown in this book, the history of the Sun is one of turmoil and change. And now we find compelling evidence that the history of our Milky Way Galaxy is one of turmoil and change. Can all this be coincidence? Surely not.

In the new picture, it is certain that our Galaxy undergoes repeated upheavals comparable to the outbursts we see in other galaxies we call Seyferts. The simplest interpretation of Clube's evidence is that the Sun, along with many millions of other stars, moves eccentrically in and out, closer and further away from the central nucleus, as well as following a circular path around the nucleus. Blasts of material are shot out from the nucleus every 100 million years or so, and these traveling gas clouds and their associated shock waves must sweep through the neighborhood of the Sun, as they first create and then destroy the short-lived spiral pattern, only for a new spiraling blast to follow behind. We can combine with this picture elements of the old idea of a fixed spiral pattern through which the Sun moves in its circular orbit. For, whether the Sun passes through a spiral shock or a spiral shock blasts outward and overtakes the Solar System, the effects on Earth must be very similar.

So we come back, again, to the evidence of dramatic upheavals on Earth some 60–70 million years ago. At this time, both the Sun and planets received a pounding from outside—a pounding which Clube now links with the most recent explosive outburst of the core, when our Galaxy became a mini-quasar. Material captured by the Sun from the expanding wave of matter was pulled onto the Sun itself, producing changes in the solar balance like those that we have come across in previous chapters. The material also rained down on the Earth as meteoritic debris, and it may well be that the Sun's present family of

167

comets is the leftover fragmentary material of this close encounter with the world beyond the Solar System. I have already mentioned the "coincidence" that the end of the era of the dinosaurs occurred at about this time; similar upheavals and mass extinctions of both flora and fauna are found at roughly 150-million-year intervals going back through the geological record, at least as far as the start of the Paleozoic era about 570 million years ago—we simply can't tell from the record in the rocks much about what happened before then, in the first 90 percent of the Earth's life. Very recently, studies of lunar samples have shown evidence that the number of meteoroids striking the Moon follows a very similar rhythm, peaking roughly every 150 million years. Clube suggests that we are midway between cosmic cataclysms, with the next big blast due in just under a hundred million years. The evidence is circumstantial —but at least it is on his side.

Look at it this way. If someone came up with a theory of repeated upheavals of the Milky Way Galaxy, but studies of the geological record and the surface of the Moon showed that nothing much had happened to the Solar System in the past thousand million years, then we'd find it hard to accept the theory. The fact that the Solar System has undergone repeated upheavals in the past 600 million years doesn't *prove* that the whole Galaxy was involved in the process, but something certainly did shake up the Solar System, and more than once!

What does this tell us about the evolution of the Sun and other stars? It certainly upsets the applecart as far as the idea of a star sitting quietly in space with its family of planets is concerned. The old picture saw stellar evolution as a simple process of nuclear burning balancing the self-gravity of the star, with nothing else to worry about. Now we see the possibility that quite frequently, on the timescale of stellar evolution, any star in the disk of a spiral galaxy like our own may be buffeted from outside, with its luminosity altered and the balance of convection inside disturbed. This must change the standard picture of stability and steady evolution of stars, but the concepts are so new that nobody has yet come to grips with all the implications.

One thing, though, does provide further food for thought and a better perspective on the Sun's real insignificance in the cosmic scheme of things. The very dramatic upheavals of 60 or more million years ago are much greater than anything that has happened in, say, the past million or few million years. We've already seen that the very recent history of the Earth and Sun might be linked with passage of the Solar System through interstellar clouds of material; now we, or rather Dr. Clube, are suggesting that something much bigger than this happens every 100 or 150 million years. Something much bigger than the minor inconveniences of an ice age, or a temporarily disturbed and flickering Sun that can be manipulated by the insistent tugging of the planets to produce rhythms and cycles of activity big enough to force the rise and fall of whole civilizations on Earth. What the Solar System has just been through doesn't even count as a proper upheaval in Clube's book, yet that insignificant encounter with a relatively minor veil of gas and dust may have been responsible for the rise of civilization as we know it, thanks to the changing environmental conditions which made cunning and intelligence such effective tools for survival. Even if you are conservative enough to reject Clube's ideas as unproven, you still have these little ripples to explain, and you still have to face up to the evidence that a much smaller disturbance than his cosmic catastrophes has upset the Sun in the very recent past.

Let's face it, timescales of hundreds of millions of years aren't really of much direct interest to us. Even a million years is incomprehensibly long to one person, or one civilization. It's all very well having this background, but what matters in our concept of "long term" is the state of the Sun now and for the next hundred or few hundred years. When we come back down to this timescale, we can see that the neutrino evidence and the patterns of the most recent ice ages, plus the long-term records of solar activity from the historical and carbon-14 records, all suggest that the Sun has been off color for quite a while. This seediness takes the form of flickering, true; but it also results from nuclear fusion being temporarily suppressed at the center, with the Sun perhaps being a little cooler than usual. Ironically, the problem for our civilization is *not* that the Sun is off color —our civilization grew up as a result of the Sun's temporary indisposition, and depends on it. The worst thing that could happen for mankind in the immediate future is that the Sun might get back to normal, warming up and stabilizing, bringing back conditions on Earth which were wonderful for dinosaurs, perhaps, or for tree-living tropical apes, but which we might be quite unequipped to cope with.

So everywhere we look in the future there is uncertainty: the threat of a Little Ice Age or something worse starting immediately; the threat of a tropical era starting—who knows when?; and the threat of a galactic convulsion within a hundred million years, perhaps doing for the mammals what the previous upheavals did for the dinosaurs. Can we cope?

CHAPTER 11

# Out of the Trap

QUITE FRANKLY, FOR A PLANET-BOUND CIVILIZATION THERE isn't much hope. But need our civilization remain planet-bound? Let's take the cosmic perspective we've built up by looking at some of these problems on the grand scale, and look at how our descendants might, after all, survive the worst the Sun can throw at us.

Already the idea of genuine space colonies in Earth orbit is taken seriously—seriously enough for Dr. Gerard O'Neill of Princeton University* to be head of a research team investigating how to get the idea, literally, off the ground. With every new setback for nuclear power, and every new oil crisis, their best proposal looks more and more attractive—build power stations in orbit to collect free solar energy which can then be beamed down to the ground by microwave radiation, punching its way through the atmosphere to receivers like overgrown radio telescopes. The fictional space shuttle we met in Chapter 1 could well be involved in such a project when the space-flying shuttle becomes fact in the early 1980s.

*and author of the fascinating book *The High Frontier;* see bibliography.

171

This is the "proper" way to open up a new frontier. Mankind, as a species, won't get into space on a secure basis just because a few pie-in-the-sky dreamers want to do it, or because scientists want to explore outer space. It can only happen if there are pressures from industry and government, commercial pressures which, say, may result from the realization that electricity from space might be cheaper and safer than electricity from nuclear plants on Earth. There are also many proposals for "factory" processes in space; while the US space effort, even allowing for the shuttle, has been running down, it is no longer news that two or four Soviet cosmonauts man their orbiting observatories for several months at a time. And what do they do up there? Obviously, a lot of Earth resources study—both for peaceful use and for providing militarily useful information about the regions they fly over. But they also do experiments on welding in space (practicing to build a *really* big space station), growing crystals in weightless conditions (which could lead to computers as far beyond the present microprocessor as the latter is beyond the valve machines of the 1940s), casting *perfectly* spherical ball bearings, and many more.

Once western industry realizes the full benefits (in hard cash terms) of space factories, the bandwagon will start rolling so fast that it will never stop. And, remember, the Space Shuttle might well become available to commercial operators, like any aircraft, taking space out of the monopoly of government (the consequences could be interesting, although it's unlikely anything as extreme as the scenario played out in the James Bond film *Moonraker* will actually come to pass!).

How far, and how fast, will that bandwagon roll? Quite literally, once the brakes are off the sky's the limit. Earth orbit, and all the valuable work of O'Neill's team, becomes pretty small potatoes once you look beyond a hundred years or so. The problem is that we may not survive even that long, either because of the flickering Sun or because of our own stupidity, but there must be a better than even chance that we will. By then, factories in space will have been followed by workers and their families; children will be born and live their entire lives without ever setting foot on Earth, and a space generation will grow up to regard the whole Solar System as home. What will they make of it?

For a start, they'll certainly have a more sensible attitude than we toward the Sun. They will come to regard it as their power source—the central fusion reactor at the heart of the Solar System—but in no way as a god. They'll have no reason to regard it as perfect and unvarying, and they will be ideally placed not only to make use of its power but to adjust to its erratic flickers. If all goes well, the spread of inhabited space cities will continue; first, they may form a ring around the Earth; then, they may spread to make a ring around the Sun, at a comfortable distance; finally, as US space expert Freeman Dyson suggested as long ago as 1966, they may form a complete sphere surrounding the Sun. It sounds crazy—it sounds like science fiction—but it is certainly possible, as Dyson has spelled out. The metal world could be constructed by taking Jupiter apart, using its material as the building blocks of what is dubbed the "Dyson Sphere." Writer Adrian Berry spells out one scenario in his book *The Next Ten Thousand Years.* * He suggests that unmanned flying fusion reactors should be sent to orbit Jupiter inside its upper atmosphere, sucking in hydrogen, converting it into heavier elements, and using the power released to keep themselves aloft. The heavy materials produced could be used to build artificial planets, or literally a sphere around the Sun, trapping its energy to

*See bibliography.

173

power the civilization occupying the Dyson Sphere. Or perhaps a ring around the Sun would be sufficient for the job —Larry Niven's fictional *Ringworld* becoming reality. In that book, Niven, as scrupulously accurate as any professional scientist, gives the dimensions of his ring, orbiting a star rather like our Sun, as ninety million miles in radius, six hundred million miles around, and a million miles across (four times the distance from Earth to the Moon) from edge to edge, with a total mass just under that of Jupiter. The total surface area is three million times that of the Earth—more than enough living space for the human race, and much more accessible than flying around the Galaxy (assuming that it ever became possible) to find three million or more "new" planets orbiting their own suns.

The point isn't that we could do it today, or even that the ideas spelled out by Dyson, or Berry, or Niven are accurate blueprints for how to do it in the future. The point is that we can see that such a concept as the Dyson Sphere, or Niven Ring, is within the bounds of physical possibility, even given our twentieth century understanding of the physical Universe. Leonardo Da Vinci could see, from the understanding of the physical Universe that he had, that flying machines ought to be feasible—but he couldn't build one himself. Given another couple of hundred years, let alone a thousand, our descendants have every prospect of starting to rebuild the Solar System, practicing, no doubt, with the asteroids between the orbits of Mars and Jupiter, asteroids valuable in their own right (many contain iron, nickel, and other heavy elements) and, perhaps, as the basis for human colonization through a ring much more tenuous than a Niven Ring, but still offering scope for population by human ingenuity (hollowed-out asteroids full of air? little mini-planets carrying domed cities?).

To such a civilization, once it had even a toehold in the Solar System beyond Earth, solar flickers on the scale that produce catastrophe down here would hardly matter at all. If you create an artificial environment, then you have control over it. If the Sun reduces its heat output by one or two percent, then you adjust your automatic systems accordingly, and carry on (assuming you've had the sense, that is, to build your systems with a large enough range of adjustment permitted). The stay-behinds on Earth may watch the glaciers spreading over the farmlands while their more adventurous brethren are unaffected; or the ground may bake in drought while the spacefarers actually welcome a slight increase in solar output as providing more energy!

Paradoxically, it may well be easier to look after such changes in the solar furnace in an artificial environment designed for the job than to try to control the complex natural environment of planet Earth with its complex and inconvenient interactions between land and sea, sea and air, water and ice. The time may come when it isn't worth keeping the old place going—the Earth too may be dismantled to be rebuilt as part of the Dyson Sphere.

Crazy ideas? They may sound crazy, but in fact they represent the conservative end of the spectrum of speculation about man's future in space. I've made no mention here of the possibility of *controlling* the Sun, just of learning to live with it and make the best use of its energy, 99.9 percent of which is just radiated away into space. The real visionaries see the prospects of our own, or other, civilizations controlling not just one star but many, and Adrian Berry says

If there is some fundamental law that says we cannot, during the course of millions of years, occupy and exploit our entire Galaxy of 100 billion suns, then that law is now hidden from us.*

*The Next Ten Thousand Years, p. 168.

175

All this implies interstellar travel—either somehow faster than light (which really means by magic), or within the laws of physics as we now know them, but for a civilization that doesn't regard ten, twenty, or a hundred years as too long to spend in getting from place to place. Certainly we need that kind of perspective if we are to have a civilization which will last for longer than the reign of the dinosaurs. Heady stuff, but we've been around for such a pitifully small length of time compared with the great successes of evolution, like those dinosaurs, and yet achieved so much in ways completely different from any other terrestrial life form, that something along these lines really could be in the cards. Every life form except for man has always been at the mercy of the environment on Earth, and completely vulnerable to flickers in the solar furnace. Man is no longer unprotected, but has learned to make himself comfortable provided the world outside doesn't change too much. A major shift in solar output of the kind that probably sufficed for the dinosaurs would still be too much for us to cope with today; in a hundred years, though millions would die, the story might be different in that the species would continue, growing again from the nucleus of the space colonies. That's how close we are to breaking out of the trap the Earth represents, a trap baited by balmy breezes and gentle sunshine which could turn, in the space of a few score years, either to freezing blizzards or the scorching heat of the desert. The visionaries usually refer to the Earth as a cradle for mankind, but that is far too cozy a view. This planet, and its relationship with a rather erratic little star we call the Sun, constitutes a trap which has already seen the death of dozens of species before, usually just when they seemed settled for eternity. Perhaps that more realistic analogy will help to encourage the idea that we ought to get out of here while the going is good. All we need is another hundred years or so, and we will have achieved the remarkable feat of breaking out of the trap in one short interglacial, an interval of 10,000 years between the jaws of ice ages millions of years long, themselves sandwiched between still longer stretches of enervating heat. I hope we make it, even though I won't be around to see mankind finally break its shackles.

APPENDIX

# Stellar Nucleosynthesis —What Cooks Inside the Sun

STARS MAKE HEAVY ELEMENTS OUT OF HYDROGEN AND HE-lium by nuclear fusion reactions in their interiors. These reactions are what makes a star, such as the Sun, hot; the process by which elements are built up is called stellar nucleosynthesis. Astrophysicists can determine how much hydrogen, helium, and other material there is in the Universe by studying the spectra of light from stars and galaxies—every element reveals its presence by characteristic spectral lines, which are stronger if more of the element is present in a particular star. These studies show that in young stars like the Sun about 25 percent of the mass is in the form of helium, 2 percent as heavier elements, and the rest as hydrogen. It is clear that the hydrogen is primeval—it dates from the very origin of the Universe in the Big Bang some 20,000 million years ago. But although hydrogen can be fused into helium, it doesn't seem possible for even 10 percent of the mass of our Galaxy to have been processed in this way since the Galaxy formed. So much energy would have been released in the process that the Galaxy would have been ripped apart; so almost certainly a good part of the helium in stars around us is also left over from the Big Bang; theoretical calculations show that in the maelstrom of radiation and hot material at the very beginning, essentially all of the helium in the Universe could have been made from the combination of neutrons and protons to form "deuterons," which in turn stick together to form helium-3 and helium-4 nuclei. All this happened in the first ten minutes of the existence of the Universe!

So the first stars ever formed contained both hydrogen and helium. Fusion reactions in their interiors built up heavier elements which were then scattered as massive stars ran quickly through their life cycles and exploded. So a relatively young star like our Sun contains some heavy elements—that 2 percent or so—to start with, and these help to establish the particular nuclear reactions that keep our Sun hot.

The simplest form of hydrogen burning occurs at temperatures around 10 million degrees, the proton-proton (p-p) chain:

$$p + p \rightarrow {}^2H + e^+ + \nu$$

where $^2H$ is a hydrogen nucleus, $e^+$ a positron, and $\nu$ a neutrino.

$$^2H + p \rightarrow {}^3He + \gamma \quad (^3He \text{ is Helium-3; } \gamma \text{ is a gamma ray})$$
$$^3He + {}^3He \rightarrow {}^4He + p + p \quad (^4He \text{ is Helium-4})$$

The net effect of this is the conversion of four protons into one helium-4 nucleus and the release of energy as gamma rays and neutrinos.

Another cycle can operate in the Sun since it already possesses some carbon (C):

$$^{12}C + p \rightarrow {}^{13}N + \gamma \quad (^{13}N \text{ is nitrogen-13})$$
$$^{13}N \rightarrow {}^{13}C + e^+ + \nu$$
$$^{13}C + p \rightarrow {}^{14}N + \gamma$$
$$^{14}N + p \rightarrow {}^{15}O + \gamma \quad (^{15}O \text{ is oxygen-15})$$

$$^{15}O \rightarrow {}^{15}N + e^+ + \nu$$
$$^{15}N + p \rightarrow {}^{12}C + {}^4He$$

Again the net effect is conversion of four protons to one helium-4 nucleus and the release of gamma rays and neutrinos. This also occurs at about 10 million degrees.

Once all the hydrogen fuel in the middle of a star like the Sun is exhausted, it contracts and heats up to 100 million degrees, where helium burning begins:

$$^4He + {}^4He + {}^4He \rightarrow {}^{12}C + \gamma$$
$$^{12}C + {}^4He \rightarrow {}^{16}O + \gamma$$

Here heavy elements begin to be built up by addition of helium-4, going up in steps of atomic number 4 at a time. Eventually, with helium fuel also exhausted, a star may begin carbon burning at temperatures around 500 million degrees, with two carbon-12 nuclei fusing to make magnesium-24 and so on. Intermediate nuclei are produced when these nuclei "decay" by ejecting a proton or neutron, so that elements like sodium-23 begin to appear in the stellar brew. Ultimately, all elements up to iron-56 can be produced by stellar nucleosynthesis; then, the power of a supernova explosion is needed to go further up the ladder to the heaviest elements of all.

# BIBLIOGRAPHY

Bailey, M. E., and S. V. M. Clube. "Recurrent Activity in Galactic Nuclei." *Nature* 275 (1978): 278.

Bandeen, W., and S. Maran. *Possible Relationships Between Solar Activity and Meteorological Phenomena.* NASA Contractor Report, NASA X-901-74-156. Washington: NASA, 1973.

Berry, Adrian. *The Next Ten Thousand Years.* London: Cape, 1974.

Blizard, J. B. *Long Range Solar Flare Prediction.* NASA Contractor Report, NASA CR 61316. Washington: NASA, 1969.

Cameron, A. G. W., and J. W. Truran. "The Supernova Trigger for Formation of the Solar System." *Icarus* 30 (1977): 447.

Challinor, R. A. "Variations in the Rate of Rotation of the Earth." *Science* 172 (1971): 1022.

Clark, David. "Our Inconstant Sun." *New Scientist* 81 (1979): 168.

———, and Richard Stephenson. "An Interpretation of the Pre-Telescopic Sunspot Records from the Orient." *Quarterly Journal of the Royal Astronomical Society* 19 (1978): 387.

Clube, S. V. M. "The Origin of Gravity." *Astrophysics and Space Science* 50 (1977): 42.

———. "Do We Need a Revolution in Astronomy?" *New Scientist* 80 (1978): 284.

Danjon, A. "Sur la variation continue de la rotation de la Terre." *Comptes Rendus Acad. Sci. Paris* 254 (1962): 2479.

———. "La rotation de la Terre et le Soleil calme." *Comptes Rendus Acad. Sci. Paris* 254 (1962): 3058.

de Jong, T., and A. Maeder, eds. *Star Formation.* Dordrecht: Reidel, 1977.

Dicke, R. H. "Is There a Chronometer Hidden Deep in the Sun?" *Nature* 276 (1978): 676.

Dingle, L. A., G. von Hoven, and P. A. Sturrock. "Test for Planetary Influences on Solar Activity." *Solar Physics* 31 (1973): 243.

Dixon, R. T. *Dynamic Astronomy.* 2nd ed. Englewood Cliffs, N.J.: Prentice Hall, 1975.

Druzhinin, I. P., and N. V. Khamyaova. *Solar Activity and Sudden Changes in the Natural Processes on Earth.* NASA Contractor Report, NASA TT F-652. Translated from the Russian. Washington: NASA, 1973.

Eddy, John. "The Maunder Minimum." *Science* 192 (1976): 1189.

———. "The Sun Since the Bronze Age," in: *Physics of Solar Planetary Environments,* ed. by D. J. Williams. American Geophysical Union, Washington, 1976.

Eddy, J. A. "Climate and the Changing Sun." *Climatic Change* 1 (1977): 173.

Ellison, M. A. *The Sun and Its Influence.* London: Routledge and Kegan Paul, 1959.

Foukal, P. V., P. E. Mack, and J. E. Vernazza. "The Effects of Sunspots and Faculae on the Solar Constant." *Astrophysical Journal* 215 (1977): 952.

Frazer, J. G. *Adonis, Attis, Osiris.* London: Macmillan, 1906.

———. *The Golden Bough.* Macmillan, 1922.

Gauquelin, Michel. *The Cosmic Clocks.* Chicago: Contemporary Books, 1974.

Gnevyshev, M. N., and A. I. OL', eds. *Effects of Solar Activity on the Earth's Atmosphere and Biosphere.* Translated by E. Vilim and English edition edited by Hilary Hardin. Jerusalem: Keter Publishing, 1977.

Goodavage, Joseph F. *Our Threatened Planet.* New York: Simon and Schuster, 1978.

Gough, Douglas. "The Shivering Sun Opens Its Heart." *New Scientist* 71 (1976): 71.

Gribbin, John. "Climate, the Earth's Rotation and Solar Variations," in: *Growth Rhythms and the History of the Earth's Rotation.* Edited by E. D. Rosenberg and S. K. Runcorn. London and New York: Wiley, 1975.

———. *Galaxy Formation.* London: Macmillan; New York: Halsted, 1976.

———. *Our Changing Universe.* London: Macmillan; New York: Dutton, 1976.

———. *White Holes.* New York: Delacorte, 1977.

———. "New Chinese Results Tie Up Sun Cycles and Earth Weather." *New Scientist* 76 (1977): 703.

———. *The Climatic Threat.* London: Fontana, 1978. Published in U.S. as *What's Wrong with Our Weather?* New York: Scribner's, 1979.

———. "The Seeds of Life." *Analog* XCVIII, no. 2 (1978): 57.

———. "Much Ado About Nothing." *New Scientist* 80 (1978): 780.

———. *Timewarps.* New York: Delacorte, 1979.

———. *Future Worlds.* London: Abacus, 1979.

———, ed. *Climatic Change.* London and New York: Cambridge University Press, 1978.

———, and Stephen Plagemann. *The Jupiter Effect.* London: Fontana, 1977; New York: Vintage, 1976.

Heath, D. F., A. J. Kreuger, and P. J. Crutzen. "Solar Proton Event: Influence on Stratospheric Ozone." *Science* 197 (1977): 886.

Hoyle, F., and N. C. Wickramasinghe, *Lifecloud.* London: Dent, 1978.

———, "Prebiotic Molecules and Interstellar Grain Clumps," *Nature* 266 (1978): 241.

Jose, P. D. "Sun's Motion and Sunspots." *Astronomical Journal* 70 (1965): 193.

Kaufmann, William. *The Cosmic Frontiers of General Relativity.* Boston: Little, Brown and Co., 1977.

————. *Exploration of the Solar System.* New York: Macmillan, 1978.

————. *Stars and Nebulas.* San Francisco: Freeman, 1978.

McCrea, W. H. "Ice Ages and the Galaxy." *Nature* 255 (1975): 607.

Menzel, D. T. *Our Sun.* Cambridge, Mass.: Harvard University Press, 1959.

Mitton, S., ed. *The Cambridge Encyclopedia of Astronomy.* London: Cape, 1977.

Nelson, J. H. "Shortwave Radio Propagation Correlation with Planetary Position." *RCA Review,* March 1951, p. 26.

————. "Planetary Position Effect on Shortwave Signal Quality." *Electrical Engineering,* May 1952, p. 421.

————. *The Propagation Wizard's Handbook.* New Hampshire: 73 Publications, 1978.

Niven, Larry. *Ringworld.* London: Victor Gollancz, 1972.

Olson, R. H., W. O. Roberts, H. D. Prince, and E. R. Hedemann. "Solar Plages and the Vorticity of the Earth's Atmosphere." *Nature* 274 (1978): 140.

O'Neill, G. K. "The Colonisation of Space." *Physics Today* 27, no. 9 (1974).

————. *The High Frontier.* London: Cape, 1977.

Öpik, Ernst J. *Climate and the Changing Sun.* Scientific American Reprint No. 835. San Francisco: W. H. Freeman, June 1958.

————. "Climatic Change in Cosmic Perspective." *Icarus* 4 (1965): 289.

Paterson, David. "A Supernova Trigger for the Solar System." *New Scientist* 79 (1978): 361.

Rees, Martin. "Galactic Nuclei and Quasars: Supermassive Black Holes?" *New Scientist* 80 (1978): 188.

Reid, G. C., I. S. A. Isaksen, T. E. Holger, and P. J. Crutzen. "Influence of Ancient Solar Proton Events on the Evolution of Life." *Nature* 259 (1976): 177.

Sakurai, Kunitomo. "Quasi-biennial Variation of the Solar Neutrino Flux and Solar Activity." *Nature* 278 (1979): 146.

Shklovskii, I. S. *Stars: Their Birth, Life and Death.* San Francisco: Freeman, 1978.

————, and Carl Sagan. *Intelligent Life in the Universe.* San Francisco: Holden-Day, 1956.

Singer, C. E. "Supernovae and Lunar Melting." *Nature* 272 (1978): 239.

Sleeper, H. P. *Planetary Resonance, Bi-Stable Oscillation Modes and Solar Activity Cycles.* NASA Contractor Report, NASA KR 2035. Washington: NASA, 1972.

Smith, R. J. "The Skylab Is Falling and Sunspots Are Behind It All." *Science* 200 (1978): 28.

*The Solar System.* Special issue of *Scientific American* 233, no. 3 (1975).

Stuiver, Minze. "On Climatic Changes." *Quaternary Research* 2 (1972): 409.

Sullivan, Walter. "Timer in the Sun's Heart Said to Control Cycle." *New York Times,* 27 December 1978, p. A15.

White, O. R., ed. *The Solar Output and Its Variations.* Boulder, Colo.: Colorado Associated University Press, 1977.

Willett, Hurd C. "The Sun as a Maker of Weather and Climate." *Technology Review,* January 1976, p. 47.

Wittmann, A. "The Sunspot Cycle Before the Maunder Minimum." *Astronomy and Astrophysics* 66 (1978): 93.

# Index

Electromagnetic radiation. *See* Magnetism

Electrons, 14ff., 48, 66–67ff., 74; and neutrinos, 98, 100; in supernovae, 79, 82

Energy, 13–21, 36–37. *See also* Nuclear reactions; specific fuels

**F**

Factories, in space, 172

Fission. *See* Nuclear reactions

Flamsteed, John, 119

Flares. *See* Solar flares

Food, 16

Forecasts, 149–56

Fossil fuels, 17–21, 36

Frazer, J. G., 21–22, 23–24

Fuels, fossil, 17–21, 36

Fusion. *See* Nuclear reactions

**G**

Gaia hypothesis, 36

Galaxies, 157–70. *See also* Milky Way Galaxy

Galileo Galilei, 1, 117

Gallium, 101

Gamma rays, 90, 162, 180, 181; black holes and, 88; supernovae and, 80

Gas, 33ff.; natural, 17, 19–20. *See also* Atmosphere; Galaxies; Stars, birth of; specific gases

Genes, 52

Glucose, 14, 15

Goodavage, Joseph, 152–54

Greenhouse effect, 32, 34ff.

**H**

Han Dynasty, 124

Heavy elements, 26, 27–28, 58, 64, 76ff., 95, 174, 179–81; and space colonization, 173. *See also* Neutrinos; specific elements

Helium, 25ff., 30, 64, 65, 70ff., 77, 82, 94, 179ff.; and birth of star, 57, 58; and neutrinos, 98, 99; supernovae and, 79

Helium flash, 71–72

Hertzprung-Russell diagram, 59, 60, 70, 71, 73

Hoyle, Sir Fred, 52–53, 166

Hydrogen, 15ff., 25, 27–28, 30, 31, 52, 57, 58, 62ff., 70ff., 77, 94, 95, 179; and birth of star, 50, 52; and neutrinos, 97–98; and space colonization, 173

**I**

Ice ages (glaciers), 38ff., 94–95, 103ff., 116ff., 138, 170

Indians, California, 22

Indium, 108

Infrared, 34, 158, 162

Ionosphere, 8

Iron, 26ff., 82, 181; in asteroids, 174; and novae, 77; and supernovae, 79